The Biology of Peace and War

Irenäus
Eibl-Eibesfeldt

TRANSLATED FROM THE GERMAN
BY ERIC MOSBACHER

The Biology of Peace and War

Men, Animals, and Aggression

THE VIKING PRESS
NEW YORK

First published in 1979 by The Viking Press
625 Madison Avenue, New York, N.Y. 10022

Published simultaneously in Canada by
Penguin Books Canada Limited

LIBRARY OF CONGRESS CATALOGING IN PUBLICATION DATA
Eibl-Eibesfeldt, Irenäus.
The biology of peace and war.
Translation of Krieg und Frieden aus
der Sicht der Verhaltensforschung.
Bibliography: p. Includes indexes.
1. Aggressiveness (Psychology)
2. Psychology, Comparative.
3. War—Psychological aspects. I. Title.
BF575.A3E313 301.6′3 77-26617
ISBN 0-670-16709-6

Printed in the United States of America
Set in Linotype Caledonia

Grateful acknowledgment is made to The Bobbs-Merrill Company, Inc., for permission to quote from Patricia Terry's translation of *Elder Edda*, Copyright © 1969 by The Bobbs-Merrill Company, Inc.

The chart on page 70 is from *Predatory Behavior of Chimpanzees*, by Geza Teleki, Copyright © 1973 by the Bucknell University Press.

Brothers will die,   slain by their brothers
Incest will break   kinship's bonds;
Woe to the world then,   wedded to whoredom,
Battle-axe and sword rule,   split shields asunder
Storm-cleft age of wolves   until the world goes down
Only hatred   in the hearts of men.

"The Sibyl's Prophecy," *Mythological Poems,*
Elder Edda

Take up your swords! Pick up your shields!
Boldly advance against cold steel!
Glory and shame now rest in your right hands;
The day brings us death or vengeance for breach of faith!

"Lay of Bjarke," *Elder Edda*

Once I was young, I traveled alone,
And I lost my way.
I rejoiced when I found a companion,
For man takes pleasure in man.

"Sayings of the High One," *Elder Edda*

Acknowledgments

My work in the field was aided by many persons and organizations to whom I wish to express my heartfelt thanks: first, to the Max Planck Society, the von Gwinner Foundation, the Thyssen Foundation, and the Deutsche Forschungsgemeinschaft; also to the German and Austrian embassies and other agencies that helped my work in every respect—their aid was often vital abroad. I thank the governments of the countries in which I worked, in particular, the governments of Australia, Botswana, Indonesia, Papua New Guinea, the Philippines, South Africa, and Venezuela. I cordially thank my friends Dr. Kuno Budack, Professor Derek Freeman, Elke and Karl-Friedrich Fuhrmeister, Dr. Inga Steinvorth-Goetz, Dr. Hans-Joachim Heinz, Dieter Heunemann, Professor Les Hiatt, Glenny Köhnke, Heide Sbrzesny, Pim Straatmans, and all the members of my team. I am especially grateful to the Christian missions, which always received me hospitably and freely gave me information, as well as to the scientific organizations of the countries I visited, which gave me helpful aid. Finally, as always, I express my gratitude to Dr. Hans Hass and Professor Konrad Lorenz for the many suggestions I have received from them in the course of the years of our friendship, as well as to my family, on whom my travels imposed many hardships.

Contents

The Biology of Peace and War

1
Is Aggression Inevitable?

Man is torn by ambivalent feelings toward his fellow man. At certain times, he seeks his companionship; at other times, he fears or even hates him.

His personality is divided: When he reflects, he is aware of the tension he lives in, and he dreams of having peace rather than facing the heroic death he often glorifies. Nevertheless, right down to our own time, the fate of nations has depended on the outcome of battles, and in the history books, wars mark, like milestones, the development of mankind. Victory has generally gone to those more skilled in warfare and better armed, and often the only trace left of the vanquished has been the smoke-blackened ruins of their settlements. It seems almost as if mankind were slotted into an escalating process of bloody selection for war. That is a disturbing thought, arousing scruples and conflicts of conscience in us that may be based on the fear of atomic self-destruction, but certainly not on that alone, for the wish for peace is older than the bomb. We hope for and aspire to world peace. But is that an attainable aspiration? Is it not chasing after a Utopia? Would not a peaceful civilization soon succumb to a less peaceful one, and does not war—by stimulating cultural and technical development, for instance—perhaps fulfill important functions without which the human race would degenerate? And finally, is our quarrelsomeness not constitutional, so that, in spite of all out efforts, ultimately it will always prevail?

In recent years, aggression and war have been the subject of much discussion, sparked off by Konrad Lorenz, who in *On Aggression* (first published in 1963) advanced the idea that aggressive behavior performs functions in the service of the preservation of the species and has been programmed into animals as the result of phylogenetic adaptations, with the result that an innate aggressive drive causes them to fight members of their own species—their conspecifics. Lorenz based this theory on innumerable observations of animals, and pointed to some striking parallels that suggest that the same applies to man.

These conclusions were vigorously disputed. Lorenz's opponents accused him of rashly and unjustifiably extrapolating findings from the animal kingdom to man; in particular, they attacked his concept of drives. They maintained that human aggression is reactive and is acquired rather than innate, since anthropological findings show that by no means all peoples are aggressive; especially, they claimed that hunters and food gatherers, who represent the earliest stage of cultural evolution still observable today, are characterized by a high degree of peacefulness, which makes it legitimate to conclude that prehistoric man also lived in peace.

Some critics further denied that the aggressive behavior observed in individual animals and men had anything to do with the phenomenon of war; they insisted that its causes do not lie in animals' or men's nature, but can be explained adequately by cultural and historical factors. Hollitscher says, for instance: "War as an institutionalized political means of applying extraeconomic pressure came into existence only after the development of the private ownership of the means of production with which antagonistic societies began"; and he also accuses ethologists of making the mistake of dealing with man as if he were an animal.[1]

Lorenz has also been criticized for explaining away aggressive behavior as "natural," thus allegedly encouraging a fatalism that puts ammunition into the hands of those who believe in the basic immutability of human society. Underlying this view is the fallacious idea that what is inborn cannot be affected by education.

Critics of the ethological approach frequently write as if ethologists were blind to the tremendous importance of cultural and environmental factors. The discussion of aggression in particular is burdened with this misconception. In a discussion on research into aggression published in 1974, Erich Fromm writes: "What could be more welcome to people . . . who are afraid and feel themselves incapable of changing the course of things leading to destruction than Konrad Lorenz's theory that violence springs from our animal nature and derives from an untamable drive to aggression."[2] But Lorenz never talked of an "untamable drive to aggression"; on the contrary, he insists that human aggression is the greatest of all dangers at the present time and that it will not be overcome if it is accepted as inevitable, as a metaphysical truth, but only if its causes are scientifically investigated. As he said in 1963:

> With humanity in its present cultural and technological situation, we have good reason to consider intra-specific aggression the greatest of all dangers. We shall not improve our chances of counteracting it if we accept it as something metaphysical and inevitable, but on the other hand, we shall perhaps succeed in finding remedies if we investigate the chain of its natural causation. Wherever man has achieved the power of voluntarily guiding a natural phenomenon in a certain direction, he has owed it to his understanding of the chain of causes which formed it. Physiology, the science concerned with the normal life processes and how they fufill their species-preserving function, forms the essential foundation for pathology, the science investigating their disturbances.[3]

Also, Lorenz has repeatedly repudiated the foolish "biologism" that sees in man "nothing but" an animal:

> Far from underestimating the difference between the behaviour of higher animals we have described and those of human achievements that are guided by reason and responsible morality, I maintain that no one is in a position so clearly to appreciate the uniqueness of these specifically human achievements as those who see them against the background of the far more primitive patterns of action and reaction that we still share with the higher animals.[4] *

* In reply to a reporter who asked when men behave like animals, Lorenz said: "Actually never. Man is comparable with animals only subject to qualifications. He is certainly not distinguished from them by having nothing whatever in common with them, for he has a great deal, but by his having acquired something essential over and above those things that he shares with them that results in a fundamental change in his behavior; and that is his intellectual function, his thought, his vocabulary, and above all his moral concerns and inhibitions. . . . The animal is unquestionably inside us and would like to come out, but it is continually held in check

I too have continually insisted that man is by nature a cultural creature. In animals, even the most minute details of behavior have often been determined by phylogenetic adaptations; man is not so rigidly determined. He therefore needs the cultural constraint for his inborn urges. True, in the last resort, his freedom of action is restricted by this cultural corset, but in the course of time, rapid changes can take place in it if changed living conditions require a new adaptation. With the aid of various cultural adaptations, man has evolved strategies for survival in the most varied environments, each of which calls for different adaptations of social behavior. The rituals for the control of aggression required by an Eskimo, for instance, are quite different from those of a Masai or a Central European townsman. Without these culturally developed patterns of behavior, orderly coexistence would be impossible. I have discussed elsewhere that man seems motivated by an urge to cultivate nearly all aspects of his life. At an early age, children at play develop rules and observe them, and nothing gives them more pleasure than playing games according to rules. This creates a sense of order and security.[5] Lorenz similarly talks of the "genuineness of culturally formed behavior" and of the fact that we regard unritualized behavior as "forbidden": we feel ashamed when we behave in an uncultural fashion. But the fact that we are programmed to be cultural beings should not mislead us into underestimating the role of the innate.

The discussion of human aggression in recent years has been marked by the prominence achieved by antibiological trends. Among the many hues on the palette of criticism, factual and philosophical arguments mingle, and the multitude of opinions obscures the progress that has been made in investigation of the subject. Where do we stand today, fifteen years after the controversy started by Konrad Lorenz?

In considering here the various arguments advanced in the light of recently established facts, my primary concern will be whether aggressive behavior is programmed into animals and men by hereditary factors and, if so, how. I shall draw on findings in the fields of animal ethology, physiology, developmental psychology, and anthro-

---

by an alert control system. And it is that constraint that makes us free, makes us human." (R. J. Humm, "Der Mann, der die Tiersprache versteht," *Die Weltwoche* [Zurich], February 3, 1958, p. 7.)

pology, and I shall examine manifestations in different cultures of aggressive behavior, its development in the young, the situations that act as releasing stimuli, as well as the various patterns of aggression control. I hope in this way to uncover both the motives that lead to aggressive conflict and those that underlie the resolution of conflict. I have collected observations on these questions on repeated trips to the Yanomami Indians of the Upper Orinoco, the Bushmen of the Kalahari (the !Ko, the G/wi and the !Kung), the Himba of South-West Africa (Namibia), various Papuan tribes (Kukukuku, Woitapmin, Medlpa, Daribi, Eipo, and Biami), the Walbiri and Pintubi in Central Australia, as well as the Balinese and others.

A comparison of the processes of phylogenetic adaptation with those of cultural adaptation reveals striking resemblances between the two that in many cases can be regarded as the result of selective pressures working in the same direction. This is especially evident if biological and cultural patterns of aggression control are compared. There are many parallels, resulting from their function, between the dueling of vertebrates and culturally ritualized competitive combat among men. I shall suggest that biological development and cultural development obey the same functional laws, and that, to a certain extent, cultural evolution thus repeats biological evolution at a higher level of the developmental spiral. If this hypothesis is correct, as I hope to show, our knowledge of biological evolution should enable us both to determine our position in cultural development and to make predictions about its further course.

This survey will show the necessity of making a sharp distinction between warfare among men and other forms of intraspecific aggression, though there are transitional forms. In the animal kingdom, intraspecific aggression seldom leads to the death of a conspecific; this is generally prevented by ritualization of the conflict. Such regulated forms of aggression also exist among men: in conflicts between members of a group, as in the animal kingdom, men are not normally destructive. Furthermore, to a noticeable extent, they are governed by biological adaptations; the course of events is guided by, among other factors, innate expressive movements until conciliatory appeals put an end to the conflict.

In contrast to this, the controls that prevent destruction are lacking in the form of aggressive conflict known to us as war, whose aim is primarily the destruction of conspecifics belonging to a different group; in the animal kingdom in general, such destruction is limited to interspecific conflicts. And warfare between different human groups does show characteristics of interspecific conflict.* Nevertheless, it fulfills functions similar to those of intraspecific aggression in the animal kingdom. It can be described as cultural adaptation in the service of the "spacing" of cultural groups.

The perfectly ritualized battles in which many vertebrates engage certainly arose out of conflicts in which harming the opponent was the objective (see page 37). In other words, intraspecific aggression in these cases was originally intended to harm. Are we entitled to conclude from this that cultural evolution might similarly lead to a ritualization of war? And that in some condition competition between groups might become superfluous? First, we must examine the functions of war. Only then shall we be able to inquire by what other means these functions could be performed.

Anyone who explores the literature on the subject will quickly note that its authors often find it hard to discuss aggression unaggressively. Sometimes this hampers exchanges between disciplines, which are nowadays more necessary than ever before. Wolf Lepenies recently wrote: "It is high time that ethologists and social scientists wasted less energy on polemics and joined in the common task of collecting bricks for building a science of man."[6] I hope that this contribution will lead to a further relaxation of frontiers and will encourage exchanges in this spirit.

* It must not, however, be equated completely with interspecific conflict in the animal kingdom; on the contrary, it is a typically human phenomenon. In the animal kingdom, interspecific intergroup collective aggression is unknown.

2

Seeking the Causes of Behavior

In the debate on human aggression, psychologists, sociologists, and anthropologists regularly discuss the work of behavioral biologists (ethologists), and the literature often gives evidence that the opponents of ethology are criticizing concepts and ideas that have long since been revised by ethologists themselves. Some write as if ethologists believed in mystic, infallible instincts; this notion of instinct has not been used in biology for a good forty years. Many misunderstandings seem to surround the ideas of heredity and adaptation. And the frequent references to the "invalidness" of applying to man conclusions drawn from animal behavior studies betrays a certain lack of familiarity with the method and theory of biology. This is not especially surprising, since ethologists themselves took a long time to clarify the matter. Decisive works, such as Wolfgang Wickler's on the significance of convergence, have only recently appeared. I shall therefore begin with a brief survey of the theories and methods of the comparative study of behavior.

Since ethology is the study of the biology of behavior, and has developed out of biology, it has introduced the methods and approach of biology into the study of behavior. Like all the other behavior sciences, ethology looks for the causes of behavior. But in inquiring why an animal behaves in a specific manner, we ethologists do not limit ourselves to studying the causative physiological mechanisms, the releasing stimuli, and the processes of individual development; we also approach the question from the point of view of selection (what purpose does this behavior serve?), and associated with that, we also inquire into its phylogenetic and historical development.

When we examine human behavior, we ask the same questions. We try to understand it both from its functional phylogenetic and developmental aspects and from its physiological-causative aspects. The discovery that in precisely definable areas of their behavior animals are programmed by phylogenetic adaptations prompted the question whether this applied to human behavior also, and recent work has shown that this is indeed the case; man too is equipped with inborn movement patterns, drives, releasers, releasing mechanisms, and learning dispositions. The parameters of cultural development are also determined by adaptations of this kind. The comparative study of rituals in different cultures, for instance, has shown that a basic similarity of structure can be demonstrated in spite of the great differences in external appearance. Feasts are structured in accordance with universal rules; their course follows a grammar that is innate in us. There are also functional laws that affect the structure of phylogenetic and cultural rituals, since the same selective pressures are at work. This means that human ethology does not restrict itself to the study of the constitutional component in human behavior, but also seeks to apply the biological approach to cultural behavior. This point is worth making, since it is occasionally asserted that human ethology is exclusively restricted to the study of the innate. When my colleague Hans Hass and I published the first research on "human ethology" in 1966, and outlined some associated concepts,[1] our interest was centered on the study of phylogenetic adaptations, and with good reason, since that was a neglected aspect of human behavior, but we had no intention of restricting the field to that aspect.

BASIC ETHOLOGICAL CONCEPTS: THE INNATE IN ANIMAL AND HUMAN BEHAVIOR

In order to lay a firm foundation for further exploration of ethological concepts, it will be fruitful to consider first the role of the innate in animal behavior and then its role in human behavior—and then to compare their functioning.

It has long been known that there are skills that animals do not have to learn. A newly hatched butterfly takes off in flight and a spider spins its first web without having to be taught. These abilities used to be ascribed to "instinct"; the same term was applied to a migratory bird setting out for its distant goal as if guided by some mysterious force. With this empty formula, questioning often ceased. No wonder, then, that the idea of instinct fell into disrepute among biologists and scientifically qualified psychologists. But all that was half a century ago.

In 1935, Konrad Lorenz published his pioneering paper "Der Kumpan in der Umwelt des Vogels" ("Companions as Factors in the Bird's Environment").[2] Building on the previous work of Oskar Heinroth and Jakob von Uexküll, he clarified what instinctual achievements really consist of. As a result of this work—and the pioneering contributions to it by Niko Tinbergen, Erich von Holst, and Karl von Frisch must also be borne in mind—it can now be stated that in precisely definable areas of their behavior animals are preprogrammed by phylogenetic adaptations. Animals in general are born with a repertoire of completely functional movements. Many, for instance, can walk immediately after emerging from the egg; others can swim or even fly. The nerve structures underlying these behaviors grow in a process of self-differentiation on the basis of developmental instructions laid down in their genome. Hence we also speak of hereditary coordinations.* These are not all always mature at birth or emergence from the egg. Some appear in the course of development, without any need of practice or a model, just as many organs develop to a fully functioning state. Pigeons, for instance, do not have to learn to fly. If they are reared in cages too narrow to let them beat their wings, their ability to fly is not in the least inhibited later. If they are released at an age at which a normally reared pigeon is well able to fly, they fly immediately. There are many other examples of this kind. No drake has to learn the courtship display; it is able to carry this out as soon as it reaches sexual maturity even if it has never seen a model.

* Often the term "fixed action pattern" is employed. However, it suggests a rigidity that does not exist. Inborn movement patterns may vary, e.g., in their intensity of performance. What remains constant is the particular pattern of muscular contraction.

An animal is able to react to specific stimuli in the environment in a meaningful manner conducive to the preservation of the species, without any training, because it is equipped with detectors that respond to these stimuli. Functionally the detectors—known as innate release mechanisms—act like filters, since they activate certain patterns of behavior only under the influence of certain external stimuli. We know that many social reactions in the animal kingdom are set in motion by these mechanisms. In such cases, in which it is important to the sender of the signal to be properly "understood," the receiver and transmitter of the stimulus are tuned in to each other in reciprocal adaptation. Striking signals—"releasers"—colors, smells, marked physical attitudes, sounds, and expressive motions (such as those carried out in the bees' dances investigated by von Frisch) are developed. These signals can often be imitated by the use of simple dummies, such as Tinbergen employed in investigating the characteristics that release fighting and courtship displays among male sticklebacks. With the arrival of the mating season, male sticklebacks occupy a territory and, at the same time, acquire a red belly and drive away red-bellied rivals. They make courtship displays to the females, which are marked by a distended silvery belly. If a male is offered an imitation stickleback faithful to the original in every respect except that it has neither a red nor a distended silvery belly, it will simply ignore it. But it will immediately attack a wax sausage with a red underside or court one with a distended silvery underside. Sticklebacks brought up in isolation behave in exactly the same way. This is only one example among many.

But animals are certainly not automata that merely respond by reflex action to external stimuli. They also act spontaneously, activated by physiological mechanisms. Ethologists speak of motivating mechanisms, or drives. "Drive" is a purely descriptive term, acknowledging the fact of internal causation. The situation is not that we are faced with mechanisms constructed according to a single pattern—far from it. Internal sensory stimuli, hormones, and factors from the central nervous system interact in a highly complex fashion to bring about readiness for a specific action.[3] Also, as Tinbergen and von Holst have pointed out, various levels of integration of drives can be distinguished, and drives such as the aggressive or eating drive can be quite differently structured in different species.[4]

There is no unitary concept of drive in nature. An important finding is that spontaneity of behavior can in many cases be traced back to the spontaneous activity of definite groups of neurons. Pioneering work in this respect was done in 1939 by von Holst, who showed that the spinal cord of fish completely isolated from extraneous stimuli was spontaneously active and sent orderly patterns of impulses to the muscular system. He separated the brain and spinal cord of eels by puncturation and then cut all the dorsal roots of the spinal cord—the means by which the creature normally receives information from the outside world and messages from its own body. On awakening from the operational shock, the artificially respirated eels began to make well-coordinated wriggling movements, proving the existence in their central nervous system of spontaneously active groups of motor nerve cells whose patterns of impulses were centrally coordinated, with the result that patterned movement orders were transmitted to the periphery.[5] The continuous discharge of these impulses is normally prevented by the intervention of inhibitory agencies. But prolonged inhibition results in an accumulation of dammed-up excitation, whose physiology is not understood. All we know is that readiness to carry out specific actions is then increased; groups of motor cells obviously have a tendency to abreact in movement.

Automatism, central coordination, and the damming up of excitation have been demonstrated in many other experiments. It has also been shown that motor learning depends on a reorganization of the relations between central automatisms, which also explains the fact that learned movements develop their own motivation, for the groups of cells underlying them are also spontaneously active.[6] In the case of hereditary coordinations, the relations between groups of automatic cells are stable. Hence the relative interval between the phases of the muscular contractions involved in the movement always remains the same, and the patterns of movement, even when they take place with different intensity, are always recognizably the same (constancy of form).

On the basis of these findings and numerous observations of his own, Lorenz suggested that energy is produced and accumulated in the central nervous system of animals and then used up in the course of definite patterns of behavior. These processes of charging up and then discharging explain why, in unchanged external conditions, there are variations in readiness for action. We know that animals that have been unable to carry out certain patterns of behavior for a long time tend to react more and more unspecifically to environmental stimuli, and that in extreme cases the behavior takes place *in vacuo* (in the void), i.e., with no discernible external cause. When the energy has been "used up," the behavior comes to an end. In this connection, Lorenz speaks of "action-specific energy," a term liable to be misunderstood. What is meant is that underlying a particular pattern of behavior there are specific, spontaneous groups of (energy-producing) neurons; it does not imply that different sorts of energy are produced.

Many neurophysiological findings tend to show that this energy model is correct. However, we know also that a particular form of behavior is normally brought to an end by means of special cutoff mechanisms, and not only because the supply of energy has been used up. Energy models and control theory models complement each other.

It has been shown experimentally that some patterns of behavior always appear in sets, while others are mutually exclusive. In many vertebrates, for instance, the flight reaction suppresses the urge to eat or fight, while aggressivity excludes courtship behavior or the tendency to swim in a school. Thus the various drive systems that underlie behavior patterns exercise inhibitory or stimulating effects on each other.

Finally, animals are equipped with innate learning abilities specific to their species. As they have to be able to modify their behavior in ways that contribute to their survival, this is not surprising. Here is a recent example. According to B. H. Skinner's classical learning theory, an animal can be made to drop any particular behavior pattern by the administration of punishment stimuli. Euler tested this theory on cocks. He punished them for all demonstrations of aggressivity; as a result, they quickly dropped all their intimidation and aggressive behavior, and they became submissive creatures, low in the pecking order. Euler then used the same method in an attempt to rid another group of cocks of submissive behavior. But this failed. These animals retained their submissive behavior when attacked by animals of higher rank, even though they received an electric shock as a punishment.[7] Obviously these creatures are so programmed that "punishment" is the releasing stimulus for submission.

Many animals have been shown to have sensitive periods in which they are especially capable of learning certain things—a bond to a partner or an object, a food preference, or a skill such as singing a song—and in many cases what has once been learned is conservatively adhered to. In these cases, we speak of imprinting.[8]

Klopfer made some remarkable observations on the physiology of imprinting. He found that a nanny goat is willing to accept its kid (or a strange kid substituted for it) only if she has been with it *immediately* after birth—even for a short period. If the kid is separated from the mother for two hours after being with her for five minutes, when it is returned to her she will accept it, while she will refuse other kids. But if her own kid is taken from her immediately after birth and returned two hours later, she will attack it as if it were a strange kid. Klopfer pointed out that immediately after birth the oxytocin-hormone level in the goat's blood is very high, but the hormone disintegrates within five minutes. If the goat's cervix is mechanically extended, as normally happens at birth, a release of the hormone takes place. In some way that has not yet been more closely investigated, this hormone must make the goat receptive to the signals given out by the newly born kid.[9]

In some quarters, the concept of the innate has been criticized on the ground that development independent of experience can never be demonstrated; that at all stages of development, even in the egg, the organism is affected by the environment. Lorenz disposed of this argument by pointing out that absolute exclusion of environmental influence is not necessary to demonstrate the innate; absolute exclusion might, in fact, spoil the demonstration, since it might result in disturbances in development by influencing processes other than the patterns of behavior under discussion. It is known that rearing animals in unrelieved darkness generally leads to degenerative phenomena in the retina, and after such "deprivation of experience," any attempt to study innate reactions to colors and shapes will be condemned to failure. The point is to withhold from the animal information relevant to the adaptation that is being considered. If it is desired to find out whether a bird song is innate or not, it is sufficient to rear the bird in total isolation from sound. If it sings the species songs in spite of that, information relevant to the song must clearly have been contained in its genes rather than having been acquired in the course of its individual development.[10]

Work by Sperry has greatly contributed to the understanding of processes of maturation from the neurophysiological angle. It used to be assumed that the nervous system was to a large extent functionally plastic, and that its "wiring" could therefore be rearranged at will. The surgery of the 1930s was based on the assumption that if an arm nerve was used to reinnervate a paralyzed leg, a functional readaptation would take place. Sperry showed that this is not so.

Functional readaptation does not take place after surgical transplantation of nerves and muscles or other end organs; the growth of nervous connections takes place highly selectively, in accordance with a development recipe contained in the genotype. Sperry transplanted a piece of skin from a frog's back to its belly, for instance, and a piece of its belly skin to its back, at an early stage of development before the nerves had grown out to the periphery. Later, when he tickled the frog on the bit of belly skin he had transplanted to its back, the frog scratched its belly.

In another experiment, he turned a frog's eyes through 180 degrees and thus reversed its field of vision. He then separated the optic nerves and destroyed the fibrils. After regeneration, normal vision was not restored; as before, the animal saw everything reversed. Obviously a strict determination is at work here, for no functional readaptation takes place. The nerve fibrils, according to Sperry, are directed by chemical stimuli to the organ for which they are chemically determined. Nowadays we can well understand how a complex nervous system grows to functional maturity. Sperry pointed out in 1971 that this was in complete agreement with the ethological concept of the innate, but added that these neurophysiological findings were just beginning to influence fields outside biology and ethology.[11]*

The idea that because ethologists demonstrate the existence of the innate they deny the possibility of influence by experience is totally mistaken.† Both reinforcing and inhibitory influences can be

* "As yet the meaning and impact of these changes has only begun to permeate into areas outside biology and ethology. In the more human areas of behavioral science like clinical psychology, psychiatry, anthropology, education, and the social sciences generally, the prevailing conceptual approach on this subject remains today essentially unchanged or very little changed from where it stood 30 years ago" ("How a Developing Brain Gets Itself Properly Wired for Adaptive Function," in *The Biopsychology of Development,* ed. E. Tobach, L. R. Aronson, and E. Shaw [New York, London, 1971], p. 32).

† Thus Roth, discussing our experiments in isolated rearing that demonstrated the development of aggressive behavior, says: "Such results show that in many animals aggressive ways of behavior obviously exist that develop 'spontaneously' without experience. But they do not prove that these cannot be influenced by experience or by learning as the case may be. The very fact that aggression in experimental animals reared in isolation can be higher than among those that grew up naturally indicates that in the social life of animals aggression is modifiable" ("Die verhaltensphysiologischen Grundlagen bei Lorenz," in *Kritik der Verhaltensforschung*, ed. idem, p. 184). Lehrman also points out that proof of the heritability of a way of behavior does not mean that it cannot be influenced

exercised. In his earliest publications, Lorenz spoke of "instinct-learning-intercalation," and the ethological work of the past twenty years has produced many examples that show how learning contributes to the integration of behavior patterns into a functional whole.[12]

Many of the observations concerning the innate in animals apply also to men. Recent work has shown that in the fields outlined above, human behavior also is preprogrammed by phylogenetic adaptations. As I have discussed that subject at length in my *Der vorprogrammierte Mensch* (Preprogrammed Man), I shall deal with it only briefly here, particularly as I shall deal comprehensively with programming as it affects human aggressive behavior. Studies of the behavior of persons born blind and deaf, and of infants, along with cross-cultural investigations, have shown that human beings are equipped with a repertoire of movements that they do not have to learn. Many expressive movements (laughing, weeping, showing anger) can be interpreted as hereditary coordinations. Moreover, like animals, human beings react to certain stimuli situations in a meaningful way that contributes to the preservation of the species, without having to be conditioned to them. Thus, to cite only a single example, fourteen-day-old infants interpreted a black spot symmetrically expanding on a screen as an object advancing toward them on a collision course and made defensive movements.[13] They innately expect tactile consequences from the visual stimulation. A good deal has been made known about learning dispositions, by, among others, Hassenstein.[14] We shall discuss the problem of the basic drives in connection with aggression.

Lorenz has described the programming that underlies our cognitive achievements.[15] The role of the innate in human behavior is, however, still underrated. Thus Horn says: "True, phylogenetically preexistent instinctual behavior still plays a part in man, as René Spitz (1965) has shown in various connections. But the remnants

by the environment ("Semantic and Conceptual Issues in the Nature-Nuture Problem," in *Development and Evolution of Behavior*, ed. L. R. Aronson, E. Tobach, D. S. Lehrman, and J. S. Rosenblatt [San Francisco, 1970], pp. 17–52). Such comments are bound to create the erroneous impression that ethologists hold a contrary view. It is, however, certain that the area within which modifications are possible is limited. Many patterns of behavior that take place with typical intensity (for example, certain courtship movements among ducks) are practically incapable of being altered by experience.

of phylogenetically preexistent behavior in man are rudimentary* and, above all, are without any survival value outside social relations."[16] This statement is simply wrong. In view of the present state of knowledge, it is, to put it mildly, astonishing that all the phylogenetic programs that affect the perception and interpretation of environmental data, as well as our motor behavior in dealing with the living and unliving world about us, should be dismissed as "rudimentary."

The enormous variety of human expressive movements may have led to the mistaken idea that the universal behavior involved is extremely modifiable. That is not so. In human facial expressions, we are in fact confronted with patterns of behavior that are to a large extent preprogrammed.

The great variety is the result of the superposition of innate and thus constant patterns of behavior that differ only in intensity. This applies to mammals in general. Lorenz demonstrated that in dog mimicry, for example, the superposition of movements showing an intention to attack or flee in various stages of intensity results in a multiplicity of facial expressions.[17] In the expression of coyness, which plays an important role in flirtation, for instance, there are simultaneous or successive movements indicating rejection and encouragement. Various innate behaviors are available for this purpose, and there are, in addition, acquired behavior patterns, such as self-concealment. But the underlying structure, based on the principle of antithesis, is always the same.

A flirting girl can make eye contact but hide the rest of her face behind her fan, or she can turn away so as to show her shoulder, and also she can alternate between looking and looking away. She has at her disposal a variety of movements indicating indifference and rejection that can be combined with others indicating liking and encouragement; she can, for instance, drop her eyes to indicate the breaking of contact while raising her brows as a ritualized expression of encouragement. Antagonistic muscles can also be activated simultaneously. A smile of friendliness can be checked so that it becomes a smile of embarrassment. For all the apparent wealth of available

* In biological terminology, "rudimentary" means disappearing because of the loss of an original function. A rudimentary structure is a relic in whose maintenance selection is no longer at work. An example is the human appendix; another is the teeth of the embryo of the whalebone whale, which never break through.

behavior patterns, they boil down to a combination and/or superposing of a few invariables in an antithetical relationship with each other.

BEHAVIOR PATTERNS INDICATING AVOIDANCE AND REJECTION	BEHAVIOR PATTERNS INDICATING READINESS FOR CONTACT
Hiding the eyes with the hand	Smiling
Hiding the lower part of the face* (equivalent to hiding a smile)	Rapid raising of the brows
Hiding the head between raised shoulders	Looking at
Dropping the head	Facing toward
Turning away the head	Turning upper part of body toward
Inclining the head sideways	Approaching
Turning away the upper part of the body (the "cold shoulder")	Nodding at
Dropping the eyes	
Looking away	
Showing the tongue	
Activating the opposites of movements indicating liking (suppressing a smile, equivalent to hiding it)	
Moving away	
Aggressive behavior patterns (playfully striking a third party, or aggression against the self such as biting nails, fingers, or lips or stamping the feet)	

Whether there are priorities that determine which antithetical elements will be combined is a question that remains to be studied. This seems to be the case. I have found that raising the brows as a

* This has the same effect as suppressing a smile by activating antagonistic muscles.

sign of liking and encouragement is generally combined with dropping the eyes. This is followed by looking away.

The rules governing the formation of definite rituals must partly have been dictated by phylogenetic adaptations.* Feast and greeting rituals, for instance, are constructed on the same basic pattern. After an opening phase, the central feature of which is the playing out of self-representation and self-appraisal antithetically combined with elements of appeasement, there is a phase of intensive communication in which the emphasis is on friendship and sympathy. During this phase, there is joint mourning for the dead, for example. Hospitality is shown to guests, mutual interest is shown, promises are exchanged; in short, many of the typical group-forming and group-reinforcing behavior patterns are displayed, even if the initial greeting is followed only by a brief conversation about the weather. The factual information is insignificant; what matters is that social information is conveyed, confirming that the channels of communication are open and that harmony prevails. Finally, before the parties separate, there is a formal farewell. This assures the continuity of the relationship. The parties express a desire to see each other again, and often a formalized exchange of gifts or merely good wishes takes place.[18] Many of the factors common to all cultures in these matters arise from the binding function of these rituals. Others are also no doubt to be interpreted as a necessary structural element (comparable to punctuation in a sentence). But the basic rules according to which these rituals are structured and many of the movements incorporated into them must surely be innate.

We humans conduct ourselves according to norms so universally distributed that we can assume that their basic rules are inborn. In

*Rituals consist of sequences of movements. The individual sequences are separated from each other like sentences in speech. Thus in the courting ritual of *tanim hed* (which I described in *Der vorprogrammierte Mensch*) several nose-rubbing movements are followed by two deep bows. The ethological approach suggests asking where this bowing comes from, what its origin was. Is it a ritualized gesture of submission? Perhaps it merely fulfills the function of separating the different items of behavior from each other, and it may have developed solely for this purpose. In other cases, the motions that appear in rituals can certainly be derived from others. The act of greeting, for instance, is introduced by a rapid raising of the brows ("eyebrow-flash") accompanied by a simultaneous upward jerk of the head. Here we have a ritualized taking cognizance of the other party. The quick raising of the brows is a derivative of the expression of surprise and is to be interpreted as a ritualized expression of pleasurable recognition ("Oh, so it's you").

response to the appearance of specific signals, for instance, innate releasing mechanisms prevent acts of aggression, and they also release altruistic behavior. They lay down the standards for how we are to react, and departures from appropriate behavior bring about appetences that are satisfied only when the normative situation is attained (on cultural norms and their relationship to the innate, see pp. 189 and 193).

THE COMPARATIVE METHOD IN ETHOLOGY

Biology is a comparative science. Whether they are dealing with organs, ways of behavior, or complex forms of organization, biologists compare characteristics. If in so doing they discover similarities, it is reasonable for them to ask how they arose. The more complicated the structures compared and the greater the resemblance between them, the more improbable it is that they are the work of chance, and the more likely it is that they were brought about by a common causative factor. Possession of a common characteristic may be attributable to a common ancestor in which it was present.

There is a widespread view that only similarities that can be traced back to a common ancestry (so-called homologies) are of interest to ethology, particularly when comparisons are made between animals and man. Actually, as Wickler has repeatedly pointed out in recent years, a special interest attaches to the similarities that different species have developed independently of each other as adaptations to similar environmental demands (analogies or convergences), because they provide vital information about the functional laws underlying the similar development of the structures in question. I can find out about the laws governing the construction of wings by studying wings of the most varied type—the transformed anterior extremities of vertebrates or folds of chitin in the insect world, or even the wing of a plane—and the more of these technical solutions I study, the more convincing my conclusions will be. Schmidt-Mummendey and Schmidt, for instance, go astray in saying: "All statements in Lorenz and Eibl-Eibesfeldt about human aggressive behavior are based on analogies between fish, geese, or wolves and man, which, uninfluenced by the findings of human psychology, do not have much value for the analysis, control, and prediction of human behavior."[19] Hollitscher is similarly misled when he writes: "Lorenz continually transcends the boundaries he theoretically fixed himself when a belief in the unity of all living things causes him to extend to man things that apply only to rats and the greylag goose. In the present state of knowledge, this is an unjustifiable incursion from one field of study to another. Even comparison with monkeys, who are closer to human beings than any other animal, is not legitimate without qualification."[20]

Critics of the comparative method should at least be aware of the basic rules of comparative morphology. The study of analogies provides information about functional laws. Anyone desiring to understand the phenomenon of pairing, for instance, would certainly be well advised to study as many species as possible from the most varied groups of animals and not confine himself to those most closely related to us.

It seems to me especially important to emphasize this point, because, in my opinion, it is not properly appreciated. When the work of ethologists is quoted nowadays in regard to the study of human behavior, it is generally in connection with its phylogenetic determinants. Unquestionably, one of the things on which our interest is certainly focused is the study of the hereditary factors in human behavior. But the study of cultural peculiarities both in the comparative study of cultures and in the comparative study of animals and man is just as important in my opinion. Man continues biological evolution in cultural evolution. His cultural adaptations are governed by laws that are in considerable part, if not mostly, the same in both cultural and biological evolution. For example, the rituals of group formation, which also have the function of setting off one group against another, show striking parallels between animals and man. The same applies to greeting and courting rituals, ritualized fights, and numerous other cases. Social structures can be compared interculturally and interspecifically regardless of whether the behavior patterns concerned are innate or are transmitted culturally.

How little the study of convergence is understood by critics of the ethological approach is shown by Schmidbauer, who writes:

> In the analysis of purely biological function, the study of convergence is thoroughly profitable, because it shows how changes in an original situation are brought about by convergent adaptation. . . . In the context of human ethology, it is to a large extent irrelevant, because here the convergences generally have different causes: those of biological evolution arise on the zoological side; those of cultural

> evolution, on the anthropological side. . . . Human ethology can be based only on homologies, or it has no firm base. The American and British ethologists who confine themselves almost exclusively to the study of the primates have correctly recognized this.[21]

Shortly before this passage, the following remarkable sentence occurs: "The study of convergence will thus never have any heuristic value if convergences among men and animals come about by different paths." But the characteristic feature of convergences is that they come about by different paths. The developmental courses of an insect's wing and a bird's wing are very different. As organs of flight, however, they are comparable, and comparison enables us to discover the special selective pressures that were at work in the formation of those structures.

The study of homologies, on the other hand, provides us with information about the common heritage that lies behind a group and thus, among other things, shows us what its potential for adaptation is. Also, it enables us to reconstruct lines of phylogenetic development.

Biologists use a number of criteria to establish the presence of homology. In most instances, formal similarity alone is insufficient. Only in the case of very complicated structures, when similarity can be demonstrated among representatives of a taxonomic group living in the most varied ecological niches, can homology be regarded as probable. One criterion is special form; another is a special position within the structural system. This makes it possible to recognize as homologous characteristics that have lost all external resemblance. There is often great variation in the shape of the cranial bones of vertebrates, for instance, but nasal or temporal bones can immediately be identified as such by their position in relation to other bones.

A highly important criterion is the linkage through intermediate forms. There are numerous fossil series that enable the development of characteristics to be seen clearly, but the comparison of living species at different levels of organization also occasionally makes such a reconstruction possible. In the study of behavior, we actually depend exclusively on comparison between recent species.[22]

FUNCTIONAL LAWS OF PHYLOGENETIC AND CULTURAL EVOLUTION

Until a few years ago, the view prevailed in the human sciences that human behavior was formed exclusively by environmental influences and that man was only culturally imprinted. Today only a few still hold that extreme view. With the clarification of ethological concepts by Lorenz and Tinbergen, the search for phylogenetic determinants in human behavior began, and in recent years, such determinants have been demonstrated.

The biological study of human behavior has not led to anything like obliteration of the differences between man and the higher mammals. True, it has been shown that the continuum of evolution extends to the field of behavior, but with increasing knowledge of animal behavior, the special nature of human behavior stands out more and more sharply against the background of the animal kingdom. Man is characterized by reason, morality, speech, and by accumulating culture, of which there are only meager traces even among our closest relatives. Thanks to the development of speech, cultural evolution to a large extent displaced biological evolution in mankind. Striking similarities, explained by the effect of similar selective conditions, are revealed by comparison of the two processes.

A term frequently used by biologists is "preservation of the species." A structure or behavior pattern is said to perform a function in the interests of the preservation of the species, which means that it is adapted, i.e., owes its existence to a selective pressure. The phrase has survived in biological terminology because it describes a state of affairs if not precisely, at any rate intelligibly. The fact that it is a static idea does not generally disturb us. We know very well from experience that a species preserves itself: rabbits produce rabbits and ducklings hatch out of ducks' eggs. Only if we follow a species' fortunes over long periods of time do we note that there is no such thing as its preservation. What is preserved is a "stream of life"[23] groping its way along countless channels and tributaries and ramifications of species formation, here getting stranded, there breaking through into a new bed and gaining in biomass, at any rate at the beginning of biological evolution.

Species in general are wax in the hands of evolution. They change, and of those that existed in the Paleozoic era, hardly any have survived unchanged into the present. There are some species that have preserved many original characteristics. The brachiopod *Lingula unguis* has apparently remained unchanged since the Paleozoic period; at least, its shell is identical with that of its fossil ancestors. But this is an exception. The rule is that species change, and change is a response to changing environmental conditions that call for new adaptations. If individuals merely reproduced themselves identically, the stream of life would eventually end up in blind alleys.

The mechanism by which species change through genetic adaptation is relatively well known to us. Undirected hereditary changes, mutations, result in a genetically conditioned variability of individuals, and it is on this variability that selection gets to work. Those able to survive can pass on their genotype to their descendants and establish themselves. The mechanism works blindly, that is, the distribution of heritable changes is undirected.

This may seem strange to many who regard the final product as perfect adaptation, as if living creatures deliberately developed toward it. It is the process of selection that creates the illusion that the process is a directed one. The undirected, groping mechanism of evolution is the only possible answer to the unpredictable environmental changes with which organisms have to cope. Only by this means is there a continual creation of those "hopeful monsters" that open up new evolutionary prospects. A phylogenetic adaptation is, as it were, the result of a game of dice. Other evolutionary mechanisms—such as a cultural evolution guided by reason—are certainly conceivable, but their success assumes a degree of knowledge that we still probably lack.

When we speak of adaptation, what we mean is that the structure we describe as adapted performs a task in the interests of the preservation of the species, and we recall the statement we made earlier that adaptation always assumes interaction between organism and environment in which aspects of extrasubjective reality are reflected. The horse's hoof, as Lorenz put it, is a reflection of the steppe: its special shape reflects certain properties of the steppe's soil.[24]

That applies to all adaptation, whether of body structure or of behavior. If an organism is to be able to adapt itself, it must first of all be in a position to inform itself about the environmental facts that the adaptation must reflect. According to the present state of our knowledge, two ways of acquiring information are at its disposal. One is through the species: Phylogenetic adaptation results from interplay between mutation and selection in which the direction of development is determined by the latter. Accumulated experiences are encoded in the genome of the species and decoded in the course of a process of self-differentiation. The other process of acquiring information results from experiences organisms are able to gather and store in their individual lives. So far as behavioral adaptations are concerned,* experiences are acquired by learning processes and stored up in the central nervous system. There are various methods of learning. The organism can gather experiences in active exploration; it can also be instructed, that is, experience can be passed on to it.

In general, organisms cling very conservatively to adaptations that have proved themselves. In this sense, those who talk of preservation of the species are not entirely wrong. Changes in species do not take place rapidly from one generation to the next. Obviously it pays to hold fast to what has proved itself and to try out new possibilities only in small mutatory steps. In the preservation of the stream of life, both a conservative principle and a progressive principle are at work.

Of all the species that have maintained themselves as agents of the stream of life, ours is the only one characterized by a new evolutionary principle, which we describe as cultural evolution. In cultural evolution, knowledge is similarly accumulated, not in the genotype but in the brain; since the development of writing, it has also been accumulated in books, and in recent times, in electronic apparatuses, as well.

Two inventions have driven cultural evolution forward. The invention of speech enabled men to pass on information to their fellows and instruct them without actually showing them what to do. Rudiments of tradition formation continually occur in the animal kingdom also. Japanese macaques, for instance, have learned to wash sweet potatoes, and they pass on the discovery from one generation to the next, though each has to be shown what to do. The invention of speech enabled men to pass on information independently of the object. They can say: "Potatoes have to be washed." Finally, with the invention of writing, the transmission of information was to a large extent made independent of the individual. Stored-up experience is available to everyone in books.

With the appearance of cultural evolution, the slow way of gathering experience through the genome was succeeded by a far more swiftly working mechanism for acquiring information and passing it on. It put at man's disposal a new adaptive mechanism that enabled him to adjust rapidly to very different living conditions and, in a cultural adaptive radiation that is analogous to its phylogenetic

* Immunity can also be acquired.

counterpart, to fill the most varied ecological niches. An Arctic hunter and food gatherer needs strategies for survival very different from those of a Bushman in the Kalahari. Cultural traditions assure the survival of different groups. Erikson coined the term "pseudo-speciation" for this process of adaptive radiation of cultures that is analogous to that of the formation of species.[25]

Another of the parallels between the functioning of culture and genome is that progress in both is the result of conflict between forces of conservation and progressive forces of change, as Lorenz observed in *Behind the Mirror*. In cultural as in biological evolution, it is exceedingly unlikely that the accumulated treasury of tested experience will become outdated from one generation to the next. Hence it is appropriate that innovative forces should be opposed by conservative ones. Cultural progress results from the balanced working of these two antagonistic mechanisms. The struggle is repeated in every generation, the progressive forces generally being represented by the young and the conservative forces by the older. We are aware of this as the generation gap. In this conflict, there must be no outright winner for, as Lorenz has pointed out, a complete breach with tradition as a consequence of a progressive victory would endanger a culture's continued existence. On the other hand, extremely conservative cultures risk defeat in competition with other cultures because of their insufficient capacity to adapt.

In our individual lives, cultural patterns of behavior provide us with a supporting skeleton that relieves us of a heavy burden: we do not have to seek solutions for every problem or continually rack our brains about the most expedient way of behaving in all circumstances. The treasury of traditional experience, tried and tested for generations, like its inherited counterparts, provides us with a certain security. Hence we speak of our beloved, familiar habits and are afraid of breaking them.[26] Perhaps it is this anxiety mechanism that reinforces the conservative element in a culture. Generally speaking, older people are more anxious. The young, in full possession of their physical and mental strength, are readier to take risks.

In cultural evolution, clinging to old habits sometimes leads to an entirely meaningless loyalty to principles of construction that have ceased to fulfill any function. Otto Koenig gives a striking example in his book on the ethology of culture.[27] He investigated the history of the cord on the headgear of Hungarian hussars. The original purpose of this cord was to secure it to the wearer's head. It lost this function but survived as an adornment, and it was retained for purely decorative purposes even when a method of securing the cap to the head was again required; for this purpose a chin strap was developed. The band on civilian hats similarly lost its original function and became purely decorative. Now there are hats with brims curving up on both sides; Koenig found a "decorative" band on them too. It was now totally functionless, completely concealed by the curving brim, a genuine rudiment, illustrating the way in which obsolete processes are stubbornly retained. A hat has to have a band.

Other resemblances between cultural and phylogenetic structures derive from the fact that both basically have to cope with similar adaptation tasks. As energy-acquiring systems (energons),[28] they survive only by virtue of a positive energy balance. Outlays on construction and maintenance, the costs of creating energy, building up reserves, and opening up markets, and other investments of that kind must never exceed income. Energy-acquiring systems are formed by selection on basically the same principles.

The parallels between culturally and phylogenetically developed rituals are striking. They serve as communication signals, and the demands made of them are in principle always the same. A signal should be as simple and unmistakable as possible. As a sign by which the species can be recognized, many damselfish (*Pomacentridae*) have developed posterlike patterns of colors that can be likened to the patterns of flags. If the signals are movements (expressive movements, symbolic actions), these are mimetically exaggerated, often rhythmically repeated, and generally carried out with typical intensity. Variable sequences of movement are simplified and fused. This applies to the zigzag dance of the stickleback and the dance of the cleaner fish, as well as to the culturally developed *tanim hed* courtship ritual of the Mount Hagen tribes in New Guinea.

Other similarities in culturally and phylogenetically developed rituals result from special functions, such as intimidation or group formation. Among man and the higher vertebrates, conspecifics provoke reactions both of hostile rejection and friendly attraction. Groups are formed and maintained across the aggression barrier in these species, and examination of bond-forming ceremonies reveals astonishing similarities. Signals of friendship which aid adults to establish and maintain contact with conspecifics have developed out of brood-care behavior patterns, in particular those of feeding, among both birds and mammals. Ritualized feeding actions in the form of beak flirtation, for instance, play an important role in the mating foreplay of many singing birds; one of the parties adopts the role of the young bird begging for food, while the other adopts that of the parent that tends it. Infantile appeals are ritualized into friendship and courtship signals. Parallel developments have taken place among mammals. Many canines, for instance, show affection by patterns of snout contact that derive from parental feeding and the appeals of the young for food. Another widespread group are rituals in which food or nest material is handed over, which has an appeasing effect. These play an important role in greeting and courtship ceremonials among birds in particular.

When we compare cultural rituals that perform a similar function, greeting rituals, for instance, we come across many analogies. Gift rituals are widespread. Guests bring gifts and receive gifts in return. They are offered food and drink or verbal good wishes. In courtship behavior infantile appeals are made and quasi-parental solicitude expressed. In these cases, the similarities in the behavior of man and other mammals derive from a shared heritage, whereas the similarities found in the courtship patterns of man, birds, and insects developed in convergence.

Among many birds, the bond between couples is maintained by synchronization rituals in which the partners act in harmony with each other. The alternate singing of various birds has been thoroughly studied. Each member of a couple sings part of a song, taking over so precisely at the moment when the other stops that the listener believes that only one bird is singing. Each bird in turn gives the other the key stimulus. Wickler suggests that the bond between the couple is brought about by the singing motive. A partner is needed if the impulse is to be lived out.

Comparable rituals of joint action are to be found in man. Many Bushman dances take place only in interplay between two groups. In the trance dance, for instance, the women sing and beat time while the men dance. The same applies to many dance games. Only those who have been initiated into the rules can join in harmoniously. These rituals reinforce the bond between members of the group and also mark it off from strangers.[31]

Further examples of analogous results of phylogenetic and cultural development are provided by the study of threat and submission behavior, which are based on the same antithetical principle in animals and man, as Darwin pointed out in 1872, in *The Expression of Emotions in Man and Animals.* One threatens by making oneself bigger—whether by raising one's hackles, wearing combs in one's hair or putting on a bearskin—and one shows submission by making oneself smaller.

3

Aggression within the Group

I equate aggression with aggressive behavior, without burdening the idea with any a priori theoretical interpretation, in agreement with Dann.[1] I distinguish between intraspecific and interspecific aggression, depending on whether the object of the aggressive behavior is or is not a conspecific. I shall discuss later the reasons for making this distinction.

Psychologists in general describe behavior as aggressive if it leads to another party's being hurt; this includes not only physical hurt (injury or destruction) but any kind of hurt, including annoyance, taunts, or insults. There are numerous specific refinements within the overall terminology. Buss[2] defines aggression as a reaction that sends out hurtful stimuli against another organism. According to Selg,[3] aggression consists of directing damaging stimuli at an organism or surrogate organism. Others add that the hurt must be intentional: if unintended hurt were described as aggressive, dentists who cause pain would have to be charged with aggression. Dollard and his colleagues, for instance, emphasize the importance of the intention when they describe aggression as a behavior sequence whose aim is to harm the person at whom it is directed;[4] similarly Berkowitz describes it as "behavior directed at the harming of an object."[5] For Graumann, "ways of behavior and trends aimed at harming enemies" are classified as aggressive.[6]

We know that human beings do act intentionally, and their intentions can be elicited by questioning, but we do not know to what extent animals are guided by conscious purposes. At most, we can state that they show appetence behavior—that they search with quite specific motivations and, in these states of readiness for specific action, respond only to quite definite environmental stimuli. Only rarely can we say with certainty that they seek these out intentionally. Also, the idea of "harm" is problematical. When two lizards engage in a wrestling bout that ends in one of them being driven away without suffering physical harm, I am not entitled to conclude without further evidence that it has suffered harm. First, I must demonstrate that its chances of survival are less than those of the lizard that drove it away. Consequently, in ethological definitions, the terms "intention" and "harm" are generally avoided, particularly since special ritualizations often prevent intraspecific aggression from becoming destructive.

What we can state objectively is that certain ways of behavior lead to animals' driving each other away or establishing ranking orders as a result of which certain privileges accrue to the winner, for example, possession of a female or priority at the feeding place. The driving away and subjection of other animals are observed effects. It could be objected that "unintended" harm done to another animal might lead to its being driven away. But that must be very unusual indeed. To drive an animal away, repeated efforts over a considerable period of time are generally necessary. An inadvertent attack on a fellow member of a group, for instance, is not repeated. Among the higher vertebrates, there are actually complicated "apologetic" behavior patterns, such as consoling and conciliating—in short, contact is fostered and not broken off.

But should we describe as aggressive all the behavior patterns that lead to the driving away or subjection of a conspecific, such as the singing by which birds proclaim their territory or the chemical signals, such as the pheromone,* which aids a queen bee to prevent the development of other queen bees? Markl has shown that among the Hymenoptera—*Polistes gallicus*, for instance—aggressivity is primarily directed against fellow members of the group, leading to a division of labor between the alpha females, who alone lay eggs and

* Pheromones are chemicals that serve as signals. They are consumed or perceived by the sensory organs of the addressee and influence behavior in a diversity of ways.

are entitled to the best food for that reason, and the lower-ranking female collectors. He states:

> The most highly developed Hymenoptera have solved this problem with special perfection: a drug, consisting of 9-oxodecenoic acid and 9-hydroxydecenoic acid, which the queen of the honey bees produces in her mendibular glands, simultaneously sterilizes the female workers—from the point of view of population genetics, an act of extreme aggression—prevents them from forming queen-bee cells, i.e., from breeding rivals to the queen, and serves as a sexual bait that lures males on her mating flight. This drug is greedily taken up by the workers, who split it among the bee populations, and it actually inhibits their reciprocal aggression.[7]

Among the crickets *Acheta domesticus* and *Gryllus pennsylvanicus*, there is a positive correlation between singing and aggression, and the former can be regarded as a ritualized form of the latter, as it suppresses aggression among conspecifics of lower rank. If the tympanum of a lower-ranking cricket is destroyed, so that it is unable to hear, the aggression is disinhibited and it attacks.[8]

Kummer defines aggression more narrowly: "By an aggressive act, I mean in the first place a massive use of physical strength against a conspecific; the function of the act is to drive the conspecific away."[9] This definition would exclude threats from the category of aggressive behavior, though most investigators include intimidation behavior and aggressive displays in the aggressive system. I think this is well founded, in the first place, because it can be demonstrated that most threatening behavior is a ritualized form of attack. To give only a few examples, a threatening animal will bare its teeth as if about to bite, it will adopt a posture of being about to pounce, or it will stamp its feet (as if advancing to the attack while remaining on the spot). And every stage of transition between threatening and fighting can be observed. Furthermore, there is the same variation in threshold values before threatening and fighting take place, which points to a common underlying physiological system. Threatening has a rather lower threshold value than fighting and accordingly generally precedes the latter. Threatening behavior accompanies fighting as an expressive movement. The play of expression of an angry individual can in itself be taken as an indicator of his aggressive mood. To restrict the definition of aggressive behavior to actual physical clashes and trials of strength and to exclude threatening behavior derived from fighting would be to make an arbitrary and highly artificial distinction.

Similarly, I can see no good reason to set up a different category for repelling mechanisms that function through the sudden intervention of powerful sensory stimuli (intimidating by bellowing, emptying of stink glands, sudden display of striking colors, e.g., of ocelli). I shall here classify as aggressive all behavior patterns that lead to the spacing out of conspecifics by means of the repelling principle or to the domination of one individual over others; consequently, I shall include behavior patterns such as territorial bird song.

Many authors insist that the term "aggression" cannot be restricted to acts having a direct physical effect, Johnson, for instance: "Most animal aggression is also neither physical nor direct, for it is usually carried out at a distance through ceremonies which involve no contact. Furthermore, if actual fighting does break out it is usually ritualized. Sometimes the 'delivery of noxious stimuli' [here the author refers to Buss's definition] involves nothing more than looking, as in the case of baboons and macaque monkeys, who may threaten an opponent simply by staring at him."[10]

Both psychologists and biologists generally include threatening behavior (staring, clenching fists, baring teeth, etc.) under the heading of aggressive action.* I shall follow this usage and base my definition on the results of the behavior. But I shall avoid the term "harm" because it is liable to encourage misleading subjective interpretations. I should also like to point out here that aggressive behavior thus defined may have developed completely analogously among different groups of animals. The aggressive behavior of vertebrates is certainly not homologous with that of invertebrates. Homologies can, however, be demonstrated among vertebrates.

A question being discussed at present is whether there are other

* I emphasize this because Schmidbauer has said that by including threatening behavior under the heading of aggression, I extend the term in a quite unusual manner, with the deliberate aim of demonstrating the universality of aggressive behavior ("obviously it is only by this means that Eibl-Eibesfeldt succeeds in sticking to his original theory of the universality of human aggression," "Territorialität und Aggression bei Jägern and Sammlern," *Anthropos* 68: 549). I shall return to this later. Schmidbauer speaks of aggression only when a conspecific suffers physical or mental damage. He calls it "paradoxical" to describe as aggressive the threatening behavior of a gorilla that results in other animals' keeping out of its way.

forms of intraspecific aggression that must be defined. Theoretically another neurophysiological mechanism, for instance, an impulse to dominate (striving for rank), might underlie an animal's tendency to drive a conspecific from its territory. But so far, no different motivation has been demonstrated, and I know of no instance in which an animal fighting for position in the ranking order behaves otherwise than in a competition for territory.

Moyer classifies the behavior in this way: aggression between males, fear-induced aggression, irritable aggression, and territorial, maternal, instrumental, and sex-related aggression.[11] But his method is inconsistent: some of his classifications are based on releasing stimuli and others on function, which is rather confusing. Further subdivisions according to function are certainly possible—I shall return to this later—but classification according to releasing stimuli seems to me to be less profitable, since the same behavior pattern can be released by different stimuli.

It is important to distinguish clearly between interspecific and intraspecific aggression. Interspecific conflict (with prey or predators) is marked, not invariably but very often, by behavior patterns that differ from those used in intraspecific conflict. A cat uses a definite display against a rival; it shows quite different behaviors when preying on a mouse. An oryx antelope will fight a lion by trying to stab it with its sharp horns; with conspecifics, it engages in a competitive fencing match without using its weapons to maim or kill.

Reis[12] distinguishes between affective and predatory aggression in the cat and contrasts some typical characteristics of the two:

AFFECTIVE AGGRESSION	PREDATORY AGGRESSION
Intense activation of the autonomous system (sympatho-adrenal)	Slight activation of the autonomous system
Adoption of threatening or defensive postures	Stalking
Making threatening sounds	Absence of threatening sounds
"Angry" attack with claws resulting in wounds to the other party	Nape of the prey's neck attacked with teeth in order to kill
Fluctuations in readiness for action (lowering of threshold)	No variation in reaction

AFFECTIVE AGGRESSION	PREDATORY AGGRESSION
Both intraspecific and interspecific	Interspecific
Often restricted to intimidation behavior	Always directed to success (killing)
Unrelated to food intake	Related to food intake
Substantially influenceable by hormones	Very slightly influenceable by hormones

This comparison shows that we are faced here with two different systems; brain-stimulation experiments provide further confirmation of this. Finally, the catecholamine neradrenaline stimulates affective aggression and simultaneously inhibits predatory aggression. Reis also classifies as affective aggression defensive behavior against enemies belonging to other species. But further differentiation is necessary; we must distinguish between an affective intraspecific agonistic system and an affective interspecific defense system.[13] (The word "agonistic"—from the Greek *agon,* competition—is used for all behavior that is functionally associated with fighting, including both attack and flight.)

Unfortunately, the necessity of distinguishing between intraspecific and interspecific aggression is often overlooked. Ardrey, following Dart, attributes the aggressivity of modern man to the predatory (carnivorous) way of life of his Australopithecine ancestors. As we know, these African ape men killed other animals with simple weapons and ate them; they were "predatory apes."[14] But stereotyped ideas about evil predators led holders of these views to overlook the innumerable vegetarians in the animal kingdom that show a large amount of intraspecific aggression; one has only to think of bulls or fighting cocks, which are the very symbol of aggression.

I shall deal now primarily with intraspecific aggression. Threatening and fighting behavior will be described as aggressive behavior, and when they appear, I shall speak of aggression and measure aggressiveness by this behavior. The term means readiness for aggression—which can vary seasonally, for instance. There are also specific and individual differences. Functionally, aggressive behavior and that of submission and flight together form an overriding unit of competitive actions that Scott has described as agonistic behavior,[15] a term that I consider well chosen.

Experiments in the brain stimulation of domestic cocks by von Holst and von Saint-Paul confirm that agonistic behavior is a single complex. A switch in behavior from aggression to flight can be brought about by sustained stimulation of a locality in the brain or an increase in the strength of the stimulus, which points to a common neuronal base.[16] By electrical stimulation of the mesencephalon and hypothalamus of cats, Hunsperger found a coherent functional system for aggressive behavior, defense, and flight.[17]

A distinction, generally associated with a value judgment condemning aggression, is often made between aggressive and defensive behavior. In practice, this distinction is hard to draw, since the behavior patterns that appear in aggression and defense are generally the same. There are, however, exceptions. An aggressive squirrel threatens by laying back its ears and grinding its teeth. On the defensive, it sticks up its ears and squeals. As this suggests the possibility of the existence of a defensive system separate from the aggressive system, I shall bear this in mind in the chapters that follow. I must repeat, however, that the distinction cannot always be made.

AGONISTIC BEHAVIOR

Fighting System:	*Flight System:*
1. Aggressive behavior Threatening Fighting	3. Submission behavior
2. Defensive behavior Threatening Fighting	4. Flight

This makes it clear that I regard aggression and defense behavior as subsystems of a fighting system. Functionally associated with it is a flight system that can be divided into the behavior patterns of submission and flight. These subsystems, in accordance with Tinbergen's Schema of the hierarchical organization of behavior, can be divided in turn into further systems. The flight system is an important counterpart to the fighting system: it prevents fighting from leading to self-destruction. "The fearless fighter does not get far," Tinbergen writes in illustration of this theme.

The agonistic system of higher vertebrates is basically organized on this fighting-flight pattern. Divergences certainly occur: a number of vertebrates lack submission behavior, and the patterns of behavior that appear in the fighting system may or may not be ritualized. If they are not, there is no clear distinction between interspecific and intraspecific aggression. There can also be variations in the strength of motivation underlying the fighting system. Depending on the strength of the fighting appetence produced by the aggressive motivation, the system may tend to be more spontaneous (aggressive) or reactive (defensive). This varies even within vertebrate families according to the particular ecology of the species concerned (p. 57).

Psychoanalysts sometimes define aggression too broadly, in my opinion. Hacker describes it as "the disposition and energy inherent in man that expresses itself originally in activity and later in various individual and collective, socially acquired and socially communicated, forms of self-maintenance ranging all the way to acts of cruelty." He starts from the original meaning of the Latin *aggredi*, "to approach," and thus equates aggression in its broader sense with action.[18]

My classification suggests hypothetical relationships between physiological systems that can be postulated on the basis of findings so far made. In many species a positive correlation has been shown between curiosity and initiative on the one hand and what biologists call aggression on the other. But this correlation implies only that animals that are at most times more aggressive are at other times more inquisitive, and not that inquisitive behavior varies with aggressivity. And this correlation certainly does not hold for all aggressive animal species. Aggression and curiosity are certainly not to be attributed to the same physiological system, even though a common factor, which might be a hormone, has a stimulating effect on both. In the case of many vertebrates, this also applies to aggression and sexual behavior, both of which—to a certain extent at a higher level of integration—are stimulated by the male sex hormone. But they are clearly distinguished at the third level of Tinbergen's hierarchy of instincts, and at that level, they have an inhibitory effect on each other.

This has not been understood by Dann, among others. As an experiment, he angered a number of individuals and tested their "performance"—by asking them to underline certain letters in a printed text, for instance—and found that in this irritated condition they were definitely less efficient. From this and other experiments, he concluded that there was no positive correlation between aggression and performance, and claimed thereby to have refuted the view that the elimination of aggression is not desirable, since it abolishes or modifies positive qualities, such as initiative, ambition, etc.[19] But only someone with little knowledge of ethology would assume that the existence of a positive correlation between aggression and performance means that an angry individual would necessarily put on a better performance. The view he claims to refute applies to the total personality: potentially aggressive persons in general show more initiative.

AGGRESSIVE BEHAVIOR IN THE ANIMAL KINGDOM

A popular misconception about wild animals is that when they fight, they fight to kill. That is true of a predator and its prey, but it is certainly not the rule in intraspecific fighting between vertebrates. However, some vertebrates show little inhibition in the matter and attack conspecifics with all the weapons at their command and kill them. European hamsters uninhibitedly attack conspecifics of the same and the opposite sex, and bite them when they intrude into their territory outside the mating period. But after a brief exchange of bites, the loser generally escapes and is not pursued; normally it suffers no further harm.[20] Conditions worthy of note prevail among some mammals living in closed groups whose members recognize each other either individually (lions, wolves) or by the possession of a common smell (rats, flying squirrels). Among these animals, aggression between members of a group is under perfect control, but against nonmembers it is totally uninhibited. Male lions even kill strange females and their young,[21] and a similar state of affairs exists among rats and flying squirrels.[22] Here we find conditions comparable to those resulting from cultural pseudospeciation in man (see p. 123).

Intraspecific fighting among vertebrates is generally ritualized, however. Ritualization can take place in various ways. Some animals have developed special defenses against assault by rivals. A fighting clownfish tries uninhibitedly to bite its opponent's sides, and the latter wards off the attack by holding out its very solidly constructed breast fins like a shield.[23] Wild boars attack each other with their tusks, and each tries to take the other's blows on its shoulders, an area protected by a thick solid layer that is called a shield. In both cases, damage to the assaulted party is prevented by an adaptation. The aggressor is not inhibited.

We often observe that animals equipped with weapons that could easily harm each other engage in competitive bouts in which none of these weapons is used in a way that could inflict damage. One of the first to point this out was Lorenz.[24] Markl has described the piranha, a predatory fish that lives in the Amazon and the Orinoco and is much feared because of its dangerous teeth, never bites other piranhas. Rivals fight by drawing up alongside each other and striking at each other with their tail fins.[25] The dueling of cichlids is more complicated. After the initial threatening stage, in which they try to intimidate from the side and front by spreading their anal fins, raising their gill covers, and sometimes opening their mouths, they agitate their tail fins, seldom touching each other but directing powerful pressure waves at the other that enable them to test each other's strength. Next, they seize each other by the mouth and try to push or pull the other away. The duel is a nonbloody trial of strength that ends with the exhaustion and surrender of one of the contestants; it folds its fins together, changes its display coloration to a less striking one, and swims away.

The males of many poisonous snakes fight at mating time, but instead of biting, they engage in a kind of wrestling bout. Rattlesnakes, for instance, raise the front third of their body and strike each other with their heads until one collapses or gives up. There are many examples of dueling among four-footed reptiles. The common lizard fights very chivalrously: A male challenges his opponent by facing it with lowered snout; his opponent seizes him by the back of the neck and holds him for a while before giving the challenger the opportunity to do the same to him. They take turns doing this until the more powerful bite or resistance of one of the parties causes the other to realize his inferiority. The latter then lies flat on his belly, strikes the ground with all four feet without moving from the spot, and finally runs away. Marine iguanas meet head on, each trying to push the other away. When one realizes it has no chance of winning, it lies flat in front of its opponent in the posture of submission. The other party respects this and refrains from further aggressive action, allowing the loser to go away. Monitor lizards wrestle standing on their hind legs. Among birds and mammals, there are many other examples of dueling.

The fighting methods of the horned or antlered ungulates are well known. Each kind of antelope has a method of its own. Some have massive rudimentary horns to ram with, others wrestle with horns so made that they hook into or engage with each other, while others fence, using the long side of their horns as sabers.[26] The horn formation plainly shows that these organs were developed primarily for intraspecific dueling. A straight stabbing horn would obviously be a better weapon against predators than the curled horn of a ram, for instance.

Sometimes a fight begins in earnest. Two dogs fighting will bite each other, but if one is no match for the other, it can show submission by a definite behavior pattern—by lying on its back, like a puppy offering itself to be cleaned by its mother. This inhibits the other party. The loser often urinates in this position and is then licked by the winner, showing that this submission behavior is an imitation of infantile behavior—in fact an infantile appeal of the kind that often brings about a switch from hostility to friendly contact. The defeated dog wags its tail while the other one cleans it and responds by wagging its tail too, and what began as violent conflict ends in friendly play.

The existence of dueling and of submission postures enables one to conclude that it is advantageous among the higher vertebrates in general for conspecifics not to kill one another. There must, after all, have been a strong selective pressure behind the development of threat rituals and complicated rules of combat. The same must apply to aggression; otherwise it would have been easy to eliminate it completely when it threatened conspecifics. Obviously this has seldom occurred. We can say that dueling developed among animals capable of killing or hunting each other in order to enable them to continue trials of strength against each other.

In many species, a duel can escalate into a battle in earnest if the defeated party cannot get away. A defeated cichlid, for instance, is battered by its opponent until it succumbs to its injuries. The consummative situation of "conspecific no longer present" must be reached (in intragroup fighting, evasive action by the conspecific is also sufficient). The inhibitions were developed to provide the conspecific with the opportunity to escape. If it is unable to do so, it is exposed to destructive attacks.

Attempts have been made to cast doubt on the validity of this rule by pointing to some vertebrates and many invertebrates that have no inhibitions against killing one another. I know too little about invertebrates to express an opinion, but so far as vertebrates are concerned, I can refute the criticism. In the first place, I have repeatedly stated that only strong selective pressure can have led to the development of a complicated fighting ritual. If an animal is able to escape quickly from its enemy, no dueling ritual will develop. When a lion uninhibitedly kills a stranger to its group, it is a notable event because it is unusual. How frequently do such killings occur? No exact figures are available. All the critics state is that there are animals in whom the inhibition on killing conspecifics is lacking. In that they are telling us nothing new.

Among many reptiles and fish, a fight can be ended if one of the combatants conceals the aggression-releasing signal, as happens when the submission posture is adopted. We have also mentioned that over and above appeasing their enemies, mammals—and birds also—are able to establish a friendly bond with them by making use of the infantile appeal. In some cases, this appeal is exercised directly by actual use of an infant, as among the Barbary apes. Males borrow a young animal when they wish to approach a senior in rank. They carry the young ape in such a way as to ensure that the senior will have a good view of it; this inhibits the latter's aggression.[27]

Aggression is often dealt with by the intervention of third parties. This applies above all to primates. Baboons of high rank tolerate no aggression among their fellows of low rank. They often side with the attacked party and drive the aggressor away.*

FUNCTIONS OF AGGRESSIVE BEHAVIOR

In general, an animal's greatest competitors are its conspecifics. They eat the same food and need the same sleeping and breeding places, which are in limited supply. If an animal population is to prosper, it must be spaced out in a way that avoids overpopulation of any particular area. By their aggressive behavior, animals exert a certain pressure on their conspecifics, enforcing their distribution over a wider area. The Central European hamster is a solitary animal; each individual occupies a definite area and fights all other hamsters that intrude into it. Robins occupy a territory in couples, and drive all intruding robins away. Other animals live in exclusive groups, whose members recognize each other by a common characteristic (group smell among rats) or individually (baboons). The

* I have discussed the phylogenetic development of mechanisms to control aggression—that of the binding ritual in particular—at length in my book *Love and Hate*, to which I refer the reader in this connection.

group jointly occupies an area and cooperates in driving away outsiders.

The area defended by an animal or group of animals is known as its territory. Territoriality is space-related intolerance.* It leads in the first place to a distribution of the animals over a wider area, including less favorable marginal zones. The inhabitants of a territory enjoy a number of advantages. They know their way about it, they are familiar with its hiding places and watering places and where food is to be found, and they know where to take refuge when danger threatens. In short, they feel secure in it.

The size of territories varies according to the supply and demand for food. An eagle needs a bigger territory than a buzzard. There are animals that have no feeding territories, because food as a limiting factor plays a smaller part. But they fight for breeding places, because there is a shortage of these. Many seabirds fight for breeding territories, and only those that succeed in winning and keeping a breeding place can breed. The others survive, but are eliminated from the reproductive process for at least one breeding season.

Fighting for territory is not constant; there are a number of rituals that prevent it. Many birds sing a song that proclaims the occupation of a territory. Woodpeckers drum on rotten branches so that they are audible over a considerable distance. Many mammals mark out their territories by using the scent of glandular secretions or of urine and excrement. Dogs, as everyone knows, use urine. The pygmy hippopotamus uses a mixture of urine and excrement. When it excretes, it waves its tail rapidly, and at the same time, the male directs a stream of urine from its backward-bent penis at the excrement. By this means, a fluid mixture of urine and excrement is scattered over the bushes of the riverbank. The creature creates a home atmosphere for itself in this way.

A bush baby that we kept urinated on its hands and then rubbed the urine on the soles of its feet. Thus when it climbed it left a striking reminder of itself on objects and walls.

Sea lions mark out a territory by making themselves as tall as possible on their boundaries and barking across to their neighbor. The latter responds in the same way. Neighbors generally respect each other's territory, and fighting is rare.

* In spring, male sticklebacks move about in shoals in shallow waters, looking for areas with water plants. When a male finds such an area, it occupies it and its behavior changes. Henceforth it attacks all male conspecifics.

A tame badger I reared many years ago marked prominent features in its territory, as well as the tips of my toe, with the secretion from the glands under the base of its tail. Hamsters use the secretion from their side glands.

Territoriality is widespread in the animal kingdom. We come across it among annelids, mollusks, arthropods, and vertebrates. That statement does not mean that we regard territoriality in the animal kingdom as invariably homologous.* Territoriality takes very different forms. There are animals that have no fixed dwelling areas, but defend the territory in which they happen to be; their territory moves about with them. Some are territorial only at certain times of the day; at other times, they admit their conspecifics without resistance. Territoriality is often limited to a definite season; among swallows and starlings, it is restricted to spring and summer.

Spatial intolerance is not always expressed by fighting. Territories can be maintained without fighting, by ritual alone. The point to note is the dominance of the occupant in his own territory. That is implied by Pitelka when he states that the real significance of territory lies not in the nature of the mechanisms by which the animal gets others to identify itself with it, but in the degree to which it is used exclusively by its occupants.[28] Wilson says a "territory is an area occupied more or less exclusively by animals or groups of animals . . . through overt defense or advertisement."[29] But when nonviolent display fails to deter an intruder, more aggressive forms of intimidation behavior, and ultimately fighting, generally ensue. According to Willis, a territory is an area in which an animal or group of animals dominates over another.[30]

Kummer defines a territory as "a field of repulsion . . . fixed in space." He points out that when territory is being established, aggression may well decide who get the best places, but as soon as the distribution has taken place, the movements of the occupants are formalized and predictable, and conflicts are reduced to a minimum. Everyone knows where the boundaries lie and respects them, at any rate for a certain time. Kummer regards the fact that territories are defended when necessary as one of the characteristics of territoriality, but points out that this defense is generally limited

* My repeated insistence that we do not automatically regard similar characteristics in different species as homologues may seem superfluous to the initiated reader. But those less familiar with the subject sometimes tend to assume that, and I wish to prevent them from falling into this trap.

to harmless demonstrations—visual or audible intimidation behavior.[31] (Examples of how a species divides up an area into territories are shown in figures 1 and 2.)

Exclusivity and Expulsion

Territorial groups are exclusive, shutting out strangers. In addition to the territorial function of spacing out, this exclusivity offers a number of advantages from the point of view of selection. Since these cannot be regarded as mere by-products of spacing out in the interests of an optimal use of space, they merit special attention.

As groups in general arise out of family units whose progeny remains together even when incest barriers are present, the members of large groups are generally characterized by a high degree of close relationship. This has effects on population genetics: When individuals in such a group sacrifice themselves for the community and thus contribute to the group's survival, they also contribute to the dissemination of their altruistic endowment, for even though their altruistic behavior excludes them from reproduction in favor of other members of the group, this altruistic endowment is con-

Figure 1. The distribution of male song sparrows (*Melospiza melodia*) on Mandarte Island in the 1961 breeding season. After Tompa (1962).

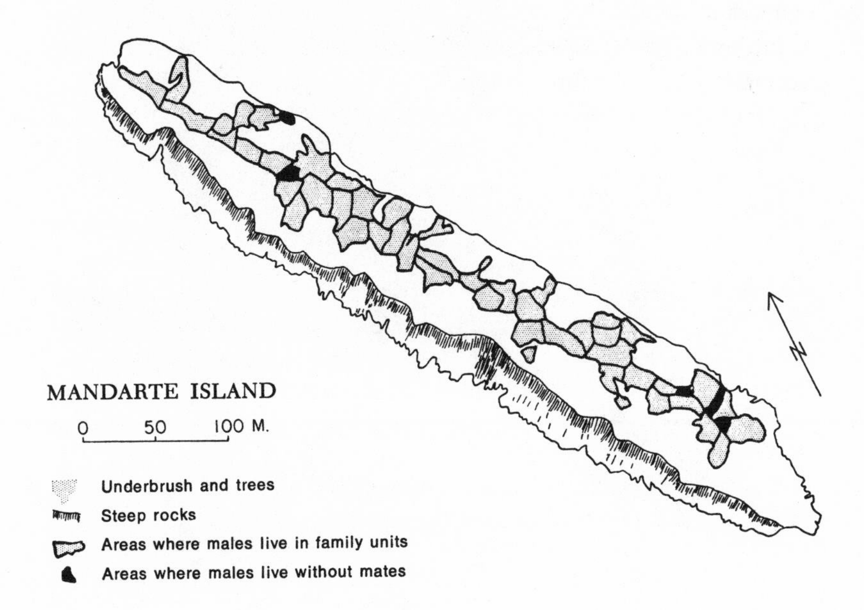

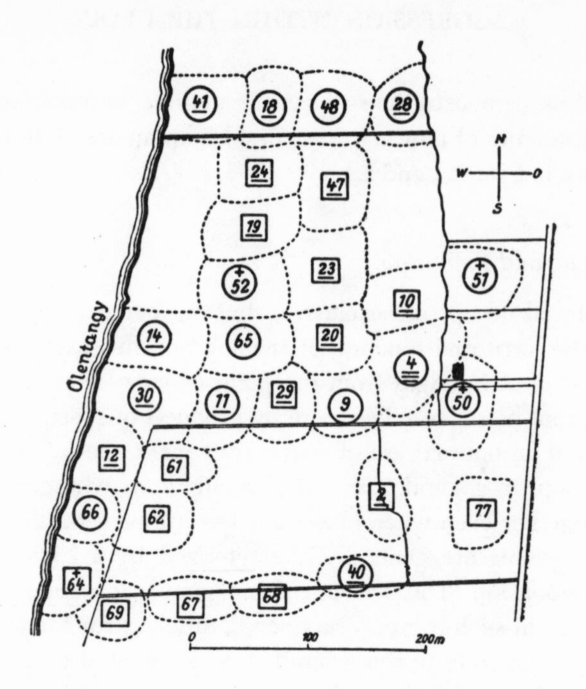

Figure 2. Territories of male song sparrows in successive years. A circle indicates a bird domiciled throughout the year; a square, a visitor in the summer only; a cross, the first year of a bird's presence; birds present the previous year are underlined; a stroke is added for each additional year. After Nice (1937).

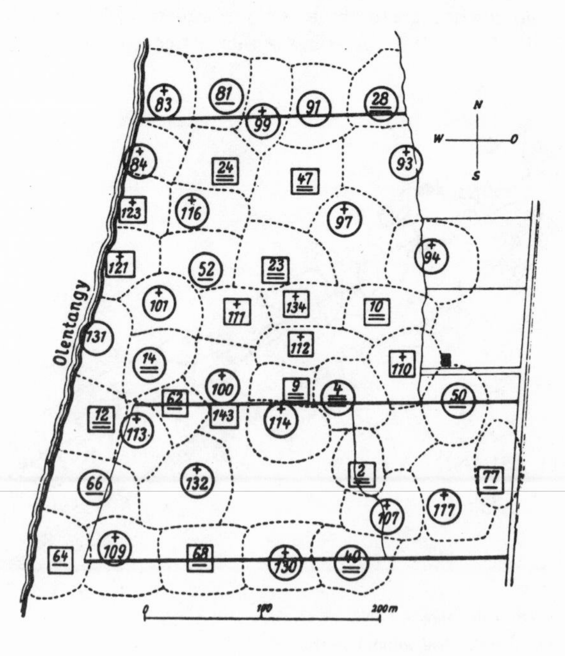

tained in the genes of members of the group closely related to them.[32]

It is only through this group relationship that the development of altruistic behavior can be understood. The smaller the group, the easier it is for mutations advantageous for selection to establish themselves. From this point of view, it is advantageous to utilize aggressive behavior to exercise exclusivity against nonmembers of the group. But exclusivity must not be absolute. An admixture of new genes into the genetic pool increases the variety of the population and provides new material for selection. Such limited exchange between groups can be observed. Among various kinds of monkeys, young males leave their own group and seek entry into others.

In man, this exclusivity of groups led to rapid cultural pseudo-speciation, involving his rapid genetic evolution.

Observation of various social animals has shown that excessive divergence in appearance or behavior from the group norm results in attack. The psychologist Schjelderup-Ebbe found that hens were attacked when he marked their combs with a colored spot or tied them in a different direction.[33] We owe a number of impressive examples of this reaction to Jane van Lawick-Goodall's recent observations of chimpanzees living in a state of nature. When infantile paralysis broke out in the group she was observing, the victims, who had previously been integral members of the group, were then avoided and attacked (see pp. 66 ff). The reaction that leads to the expulsion of sick animals certainly serves to maintain the group's homogeneity.

The same is true of human beings, who use certain forms of teasing and mockery to ensure that divergent individuals either adapt themselves to the group norm or are excluded. We can therefore conjecture that maintenance of the group norm was adaptive, at any rate for life in primitive small groups.

Sexual Rivalry

The fact that many male vertebrates fight only at mating time and solely for the possession of a female, and do not take part in rearing the young after mating has taken place, points to the existence of a selective advantage associated with fighting between rival males; the strongest and most dexterous, and also the healthiest, are selected for reproduction in this way, and to a certain extent, the species is kept healthy by this process. This seems to be the only purpose of fighting among many reptiles, and we therefore speak of mating fights. When males take part in looking after the brood, it is naturally of special importance that the best should be selected for this purpose, and fighting between rival males sees to this.

Ranking Order

Aggressive clashes are often to be observed within a social group, in which competition is by no means totally suppressed. Fighting takes place for priority at the feeding place and for other privileges, and a ranking order develops out of this fighting. Every member of the group learns from victory and defeat who his superiors and inferiors are, and behaves accordingly. Once a ranking order has been established, a mere brief threat from a superior is generally sufficient to keep an inferior in his place. Fighting is thereby avoided and order maintained; the ranking order heads off aggression within the group. This indicates that aggression did not develop for the purpose of establishing ranking order, but that the ranking order developed as a mechanism for controlling aggression that was advantageous to the group in other respects. Aggression against fellow members of a group means that the stronger gains certain advantages, such as the possession of females, and it often leads to a division of labor, with the result that it also fulfills other secondary functions that serve the preservation of the group. High-ranking monkeys are the "focus of attention" of all other members of the group. Attention is paid to what they do, and they generally also put themselves on show.[34]

According to Hold, the frequency with which a member of a group of children is the center of attention is a good criterion of his or her place in the ranking order. The author counted the number of times a child was looked at to determine its rank. Those who were most often the object of attention were also those who showed the most initiative, in Hold's observations. They decided what the others should do and were more aggressive than the average, though they were never extremely aggressive. They protected children of lower rank, settled disputes, played more often with different children, often took part in role-playing games, and were less locality bound. They also initiated physical contact more often. If they were given candy to hand out, they kept control of the other children, while low-ranking children promptly lost control of the others.

Low-ranking children copied their high-ranking fellows, obeyed them, asked them questions, and sought contact with them by offering them candy and help, and told them and showed them things. Sometimes they kept out of their way. They sought out the nursery-school teacher more often and played alone more frequently than did the high-ranking children.

High-ranking order is shown by, among other things, familiarity with the locality, for high-ranking children generally have the longest nursery-school experience.[35]

ADAPTATIONS THAT DETERMINE AGGRESSIVE BEHAVIOR

I have repeatedly pointed out that it is meaningless to ask whether complex behavior patterns are innate or acquired; we must expect that both heredity and environment were involved in their development. Consequently, no ethologist today would ask whether "aggression" is "innate" or "instinctive." Those who argue as if ethologists pose such alternatives have obviously either not read or not understood the recent literature on the subject.

As I stated at the outset, we must be more discriminating; we must ask whether there are phylogenetic adaptations that determine aggressive behavior, and if so, how. Are we confronted with these adaptations in the form of motor patterns, drives, and/or releasing mechanisms, for instance, and what part does individual experience play in the integration of possible innate components into a functional whole?[36] The anthropologist Freeman adopts the same position. "An interactionist approach to the study of aggression," he writes, "leads to the general conclusion that aggressive behaviour is determined by both internal and external variables and that it is strongly affected by learning."[37] Unlike the interactionists among the behaviorists whom I criticized at the outset, Freeman shares my belief that the contributions of heredity and environment are perfectly capable of being investigated.

Though I have elsewhere discussed how phylogenetic adaptations play their part in determining animal behavior, I shall briefly review the subject by giving a few examples:

A large number of experiments in the isolated rearing of vertebrates and invertebrates has shown that the motions with which animals fight their conspecifics are hereditary coordinations. Siamese fighting fish, sticklebacks, and cichlids develop their respective specific repertoires of fighting movements when reared in total isolation. When marine iguanas reared in isolation were given companions, they immediately attacked them by butting them with their heads; and Galápagos lava lizards brought up in the same conditions developed their specific form of combat by blows of the tail. Fighting cocks brought up in isolation fight with their spurs, and turkey cocks similarly reared wrestle with each other just as normal turkey cocks do. Since these animals develop different behavior under similar environmental conditions, the differences must be genetically based.

Characteristic fighting behavior often develops before completion of the growth of the organ with which the fighting is carried out. Lorenz vividly describes how a gosling practices striking with the bend of its still tiny wing. Horned ungulates often begin butting with their heads in their specific fashion before their horns have developed. The list of examples could be multiplied considerably.[38] Today it can be safely stated that not only the organs with which they fight their kind (antlers, horns, tusks, etc.) but also the movements that go with them are innate in many animals.

Innate Releasing Mechanisms and Releasers

We know from a number of studies that animals sometimes respond aggressively, as if by reflex, to certain signals of their conspecifics. Lack's experiments with dummies, for instance, are well known. He showed that a tuft of red feathers was sufficient to trigger an attack by robins, while a stuffed bird from which the red feathers had been removed had no effect, though it looked much more like a robin.[39] The signal that causes the pine lizard (*Sceloporus undulatus*) to fight is the blue stripes that adorn the sides of the male. If the gray sides of a female are painted blue, it is attacked just like a male; if the sides of a male are painted gray, it is courted as if it were a female. Male sticklebacks, as we have seen, are affected by similar releasers. Olfactory signals play an important role among mammals, but they have been less closely investigated. It is noteworthy that among some mammals no aggression worthy of the name is roused by members of the same group, while nonmembers are the object of violent attacks. Experiments with brown rats and domestic mice have shown that the aggression-releasing signals are obviously masked by a common group scent. If a domestic mouse is put into a strange group, it is immediately attacked by all the members of the group, but if it has been rubbed with the urine of males belonging to the group, it is accepted. Among higher mammals, it is often personal knowledge of the individual that has the protective effect.[40]

There are also signals that inhibit aggression. In the case of the cichlid *Haplochromis burtoni*, for instance, aggression is inhibited by orange-red spots on the side of the body, while a vertical black stripe on its face increases it. If a blind young fish is put into a tank as a permanent object of aggression, and an adult fish in the tank is shown a dummy with nothing on it but an orange-red spot, its biting rate directed against the blind fish declines by 1.77 bites a minute in comparison with the starting rate, while a dummy with a vertical stripe increases it by 2.79 bites a minute. When dummies marked with both are used, the biting rate increases by 1.08 bites a minute. The effect of the two signals is thus cumulative (figure 3).[41]

Learning Dispositions

So long as no harm is suffered, fighting and threatening are obviously pleasurable, since the opportunity of engaging in them can be used as an incentive for training purposes. Siamese fighting fish and cocks can be taught to carry out certain tasks if they are rewarded by being given the opportunity to threaten a conspecfic or a dummy.[42] Domestic mice learn to make their way through a labyrinth if they are rewarded by being permitted to fight another mouse.[43] These experiments show the existence of learning dispositions that reinforce aggressive behavior without further incentive.* Moreover, aggression is certainly reinforced when it leads to a definite objective, for example, priority at the feeding place. The old view that it is reinforced only by such instrumental means is, however, refuted by the above-mentioned experiments.

Rats, monkeys, and mammals have been trained to give themselves electrical stimuli by pressing a lever connected to electrodes

* Domestic mice are generally more aggressive after such successful combat (K. and K. Y. H. Lagerspetz, "Changes in the Aggressiveness of Mice Resulting from Selective Breeding, Learning, and Social Isolation," *Scandinavian Journal of Psychology* 12 [1971]: 214–18; K. Lagerspetz, "Interrelations in Aggression Research: A Synthetic Overview," *Psykologiska Rapporter* 4 [Åbo, Finland, 1974]; J. P. Scott, *Aggression* [Chicago, 1958]). On the other hand, mice that are punished for aggression become peaceful.

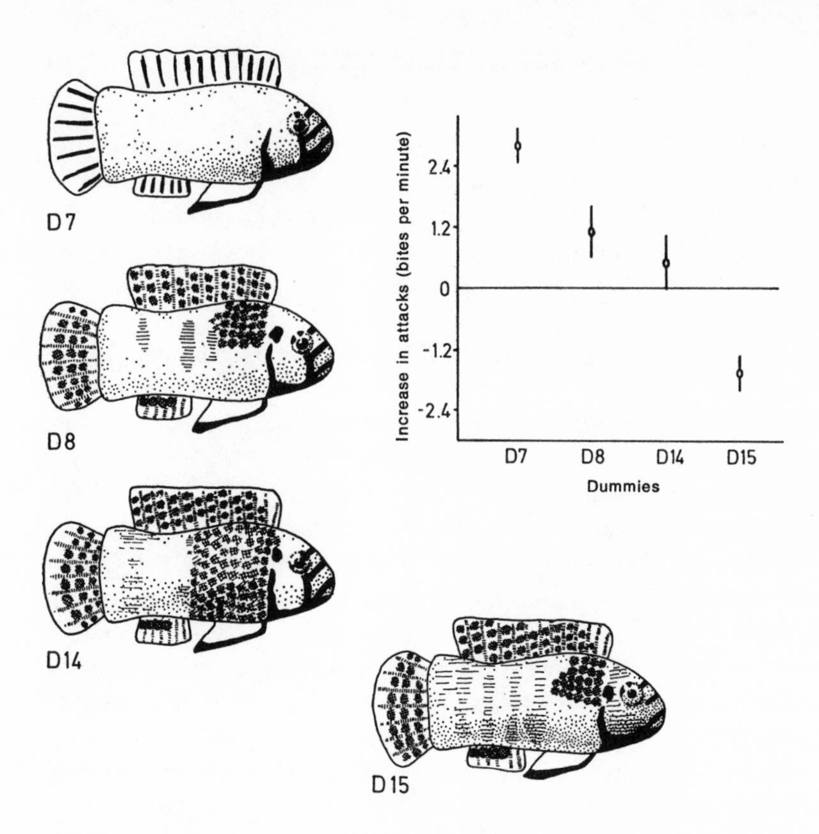

Figure 3. In the cichlid *Haplochromis burtoni*, a vertical mark on the head is an aggression-releasing signal, while orange-red spots on the side inhibit aggression. When a fish is shown dummies with a vertical head mark but without orange-red spots (D7), its aggressivity (measured by bites directed at a blind fish put into the water with it) is increased to a definite extent. A dummy without a vertical stroke but with orange-red spots (D15) diminishes its aggressivity. From Leong (1969).

implanted in their brain. This activates definite behavior patterns, and the self-stimulation is sought or avoided, depending on the position of the electrode. When the electrode is in the position in which it stimulates aggressive behavior (such as aggressive threatening), the animals always continue to stimulate themselves. But they avoid activating the flight system, and actually learn to switch off the stimulus that activates flight when it is applied to them. Also anger and aggressivity are accompanied by slow theta waves of high amplitude from the region of the hippocampus, just as positive, "friendly," reactions are.*

* "According to the hippocampal indicator, rage and attack are to be interpreted as positive emotional reactions that arise as liberations from inhibition.

Drives

If variations are observed in an animal's specific readiness to respond to certain stimuli in the environment and these cannot be correlated with changes in the environment, we are entitled to attribute them to processes taking place in the organism itself. We talk of motivating mechanisms or drives, and of sex, hunting, or aggressive drives, depending on the appetence concerned. These terms merely correspond to the fact that there are mechanisms in the animal that result in readiness for specific kinds of action. The mechanisms themselves can be of very different kinds.

Among vertebrates, hormones play a big part in the buildup of readiness for aggressive behavior. Aggressivity is increased by the male sex hormone. This explains why reptiles, birds, and mammals become so quarrelsome at mating time. Aggressive behavior can be activated at very early stages of development by the administration of hormones: in one experiment, turkey chicks started fighting after testosterone was administered to them. The female sex hormone estrogen curbs aggressivity, as experiments on rodents have shown, while progesterone increases readiness to protect the young.

Combativeness can be released by electric stimulation of the brain, as von Holst and von Saint-Paul demonstrated with fowl[44] and Delgado with rhesus monkeys and gibbons.[45] It was not necessarily motor activity that was activated in all these cases, but the urge to fight. If no opponent was present, the urge was not fulfilled. Under the influence of the brain stimulus, rhesus monkeys of high rank sought out males of lower rank as their victims and always left females alone. Individuals of lower rank never attacked their seniors, and if no victim of lower rank was available, they remained passive. Gibbons attacked conspecifics when definite areas of the brain were stimulated, but only if kept in a communal cage. If the same animals were stimulated in the open, they threatened, but did not attack.

Plotnik has pointed out that in many of these experiments, there was no effort to ascertain whether the brain stimulus caused pain

---

This interpretation is buttressed by the observation that rage and attack have a typical motor pattern of approach (in contrast to withdrawal in fear and defense). Thus, it seems less paradoxical that self-stimulation and anger go together." (E. Grastyan, "Emotion," in *Encyclopaedia Britannica,* 15th ed. *Macropaedia,* 6:765).

or displeasure, which might have produced aggression by this secondary route, as it were.[46] More attention must certainly be paid to this point in future, though it is clear from existing accounts that the animals showed no signs of pain. I was often present at the brain-stimulus experiments carried out by von Holst and von Saint-Paul, and I cannot recall anything in the behavior of the stimulated animal that suggested that its aggressivity was a consequence of painful stimuli. Also, the tame experimental animals were completely mobile and never showed reactions of avoidance before an experiment was carried out. The strict criteria that Plotnik suggested were applied to a monkey and a human being. The latter said that he felt no pain whatever in such an aggression-activating brain-stimulation experiment—only uncontrollable anger. Unfortunately, other human subjects were not asked, but they would surely have mentioned it if they had in fact felt pain.

We know that neurogenic* attacks of rage take place among human beings, caused by the spontaneous activity of groups of cells in the temperal lobes and amygdaloid nuclei. Among some animals, the aggressive drive is so strong that in the absence of opportunities to fight, they attack surrogate objects. Fighting cocks reared in isolation by Kruijt attacked their own shadows. They also tried to attack their own tails with beak and spurs, agitatedly spinning around in circles as they did so. It looked as if a quantity of accumulated excitation had to be discharged.[47] Lorenz has described these conditions in a cybernetic model, and assumed that excitation processes in the central nervous system underlie the appetence to attack. In the view of neurophysiologists the model is soundly based. Moyer, for instance, writes:

> Lorenz's hydraulic model for aggressive behavior trends is based on a number of physiological facts. If the neural systems for aggressive behavior are sensitized by changes in the chemical composition of the blood so that they transmit excitation, the "pressure" in the direction of aggressive behavior can mount. It follows that the individual is more and more inclined to display hostile behavior. But the idea that this pressure to aggression can be prevented only by showing aggression is rather too simple.[48]

Thus various kinds of mechanisms bring about combativeness. In a number of species, it leads to a less and less selective response to external stimuli, and the pressure causes the animal to respond

* Neurogenic: created within the nervous system.

to stimuli that normally would not activate fighting behavior at all. Rasa kept males of the orange chromid (*Etroplus maculatus*) in isolation. When he put females in with them, the males attacked them. Only when they were given "whipping boys" to divert their excitation was it possible to get the males, in whom a large amount of aggression was obviously stored up, to mate. If the "whipping boy" was removed, after a short interval the males attacked the females again, though they had been living with them peacefully.[49]

These cichlids are certainly one of the species in which there is little external difference between the sexes. Thus, as Wickler suggests, it is quite possible that the male is permanently aggressively excited by the sight of the female.[50]* Normally it would be able to abreact this excitation against male rivals, but since they are absent when the male is kept in isolation with the female, the constantly activated aggression ends by being turned against the latter. This possibility must certainly be borne in mind, but another experiment by Rasa with damselfish (*Microspathodon*) is more convincing. A favorable characteristic of these fish is that even their young are territorial, and one can experiment with them without any disturbing influence from a sexually motivated appetence to seek out a partner. They learned the task of finding their way through a labyrinth to a compartment from which they could see a rival whom they were allowed to fight. The longer they were kept in isolation beforehand, the longer they stayed in the compartment. Only rivals had this effect; if another object were offered them, it failed to entice them.[51] Wickler pointed out that they might possibly have so closely associated the entrance to the labyrinth with the enemy fish visible behind the destination compartment that the sight of the entrance alone might have been sufficient to stimulate the aggression,[52] but this was invalidated by the fact that the fish showed no sign of aggressive excitation at the sight of the entrance of the labyrinth (figure 4). Young males of the cichlid *Haplochromis burtoni* show a distinct increase in aggressivity after social isolation, but their sexuality remains at the same level. Only with adult fish do aggressivity and sexuality vary similarly as a result of social isolation (figure 5).[53]

Male swordtail minnows (*Xiphophorus helleri*) fought signifi-

* Reyer has since shown this to be the case. (H.-U. Reyer, "Ursachen und Konsequenzen von Aggressivität bei Etroplus maculatus [Cichlidae, Pisces]: Ein Beitrag zum Triebproblem," *Zeitschrift für Tierpsychologie* 39 [1975]: 415–54.)

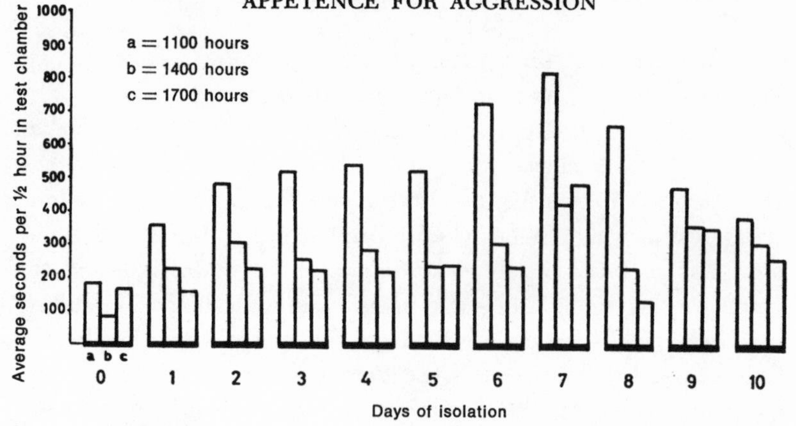

Figure 4. The increase in time spent in the test chamber after periods of isolation of different lengths. After 0, 1, 2, 3, etc., days of isolation, the same 5 fish were tested with a dummy through the test-chamber window. The increase in time spent in the chamber after increased isolation is clear. On each day of the experiment, Rasa tested the fish 3 times at different times of day. There was a marked diminution in the time spent in the chamber on the second and third occasions. From Rasa (1971).

cantly longer if kept in isolation for fourteen days (54.2 minutes as against 27.2 minutes for conspecifics not isolated). The number of threatening and butting movements per contest remained roughly the same in both groups, while the circling and mouth-fighting behavior patterns increased after isolation.[54]

The hermit crab (*Pagurus samuelis*) fights more vigorously after social isolation, though not all the aggressive behavior patterns seem to escalate in the same way. Aggressive behavior patterns classified as lower level remained unchanged, while those of high level ("actual combat") notably increased. As locomotor activity declined with prolongation of the isolation, the increased fighting cannot be attributed to the greater chances of an encounter resulting from increased mobility, but must be the direct result of accumulated aggression.[55]

The aggressivity of domestic mice increases with prolonged isolation (figure 6).[56] Physiological investigation of this very widespread increase in aggression after isolation points to its causation by, among other things, an increased concentration of catecholamtues, which act as transmitting agents to the synapses and make them more permeable, as it were.

According to Cairns, mice that have been kept in isolation gen-

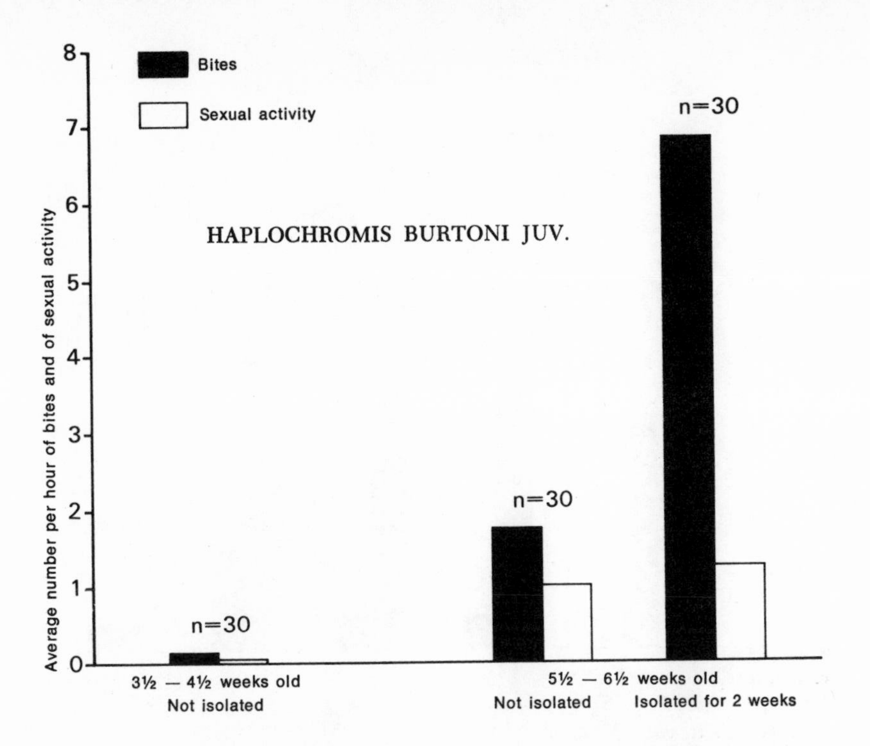

Figure 5. The increase in aggressivity among young *Haplochromis burtoni* males after temporary social isolation. Readiness for sexual activity remained unchanged, however, which shows that, at this age, the two are to be attributed to different motivational systems. From Goldenbogen (1974).

erally show increased activity and a greater tendency to investigate an unknown social partner. They sniff and touch the latter on their own initiative but, at the same time, are very timid and start back or freeze when the other animal moves. A part in this may be played by the animal's being caught up in a cyclical process escalating on the aggressive side (biting to intimidate). But the fact that the encounters always take this course and that the aggressivity of isolated animals is so great also points to a specific motivation. Even when the partner remains passive and "peaceful," the isolated mice attack without having been provoked in 56 percent of the cases, while the figure for socially experienced mice is only 19 percent.[57]*

Not every kind of animal reacts to isolation in this same way, by an increase in aggressive readiness. Among cichlids, there are, for

* In these experiments, a rat that had been reared with mice and behaved in thoroughly "friendly" fashion toward them was used as the fighting partner. This was an experimentally produced interspecific encounter. But the two species are closely related, so that intraspecific aggression was activated in the mouse.

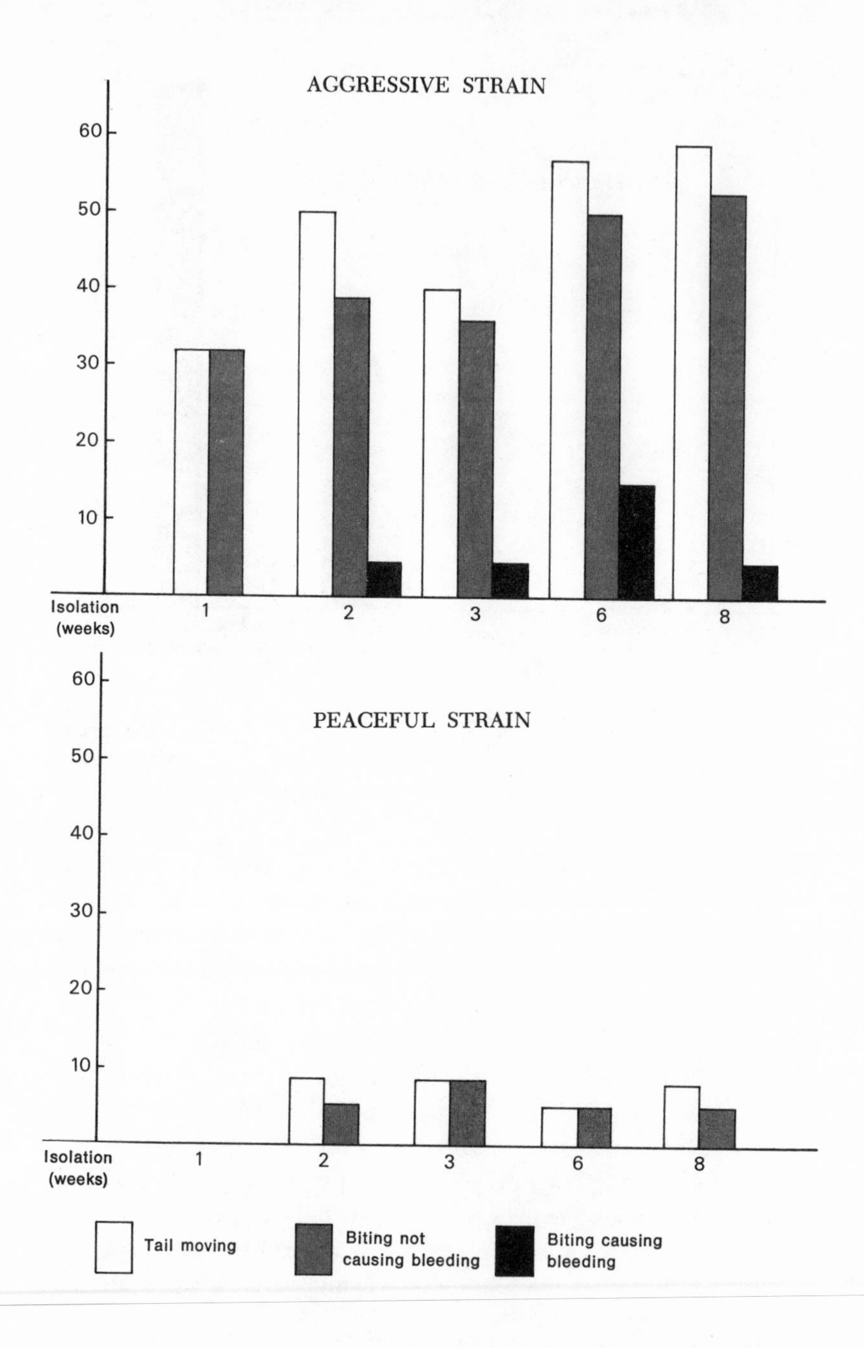

Figure 6. The increase in aggressivity after social isolation among mice of an aggressive and of a peaceful strain. The figures show percentage of males that moved their tails (white), bit without causing bleeding (gray), or bit causing bleeding (black). From Lagerspetz (1974).

example, some whose fighting drive decreases in isolation.[58] The presence of conspecifics increases their aggressivity (figure 7).

In recent years, there has been a great deal of discussion about aggression, and it has continually been argued that aggression is inherently purely reactive. Often this line has been taken for obviously philosophical reasons, but there have been other considerations as well. It has been maintained, for instance, that it is purposeless for an animal's combativeness to increase, causing it eventually to set out in search of an enemy like a knight errant. But it does not follow from the concept of the aggressive drive that storing up of aggression necessarily leads to this. A strong tie to the animal's territory would have the opposite effect.

According to Wickler, aggression can fulfill its function even without spontaneity: "Although it is advantageous for an individual or a pair to keep rivals away by aggression and thus to be able to rear young undisturbed, for instance, it would be just as disadvantageous if, after a considerable period free of disturbance, this individual or

Figure 7. The dependence of aggressivity on external stimuli in the cichlid *Haplochromis burtoni.* If a fish is repeatedly shown dummies over a period of 10 days, its aggressivity, measured by bites directed at young fish, increases to a notable extent. From Heiligenberg and Kramer (1972).

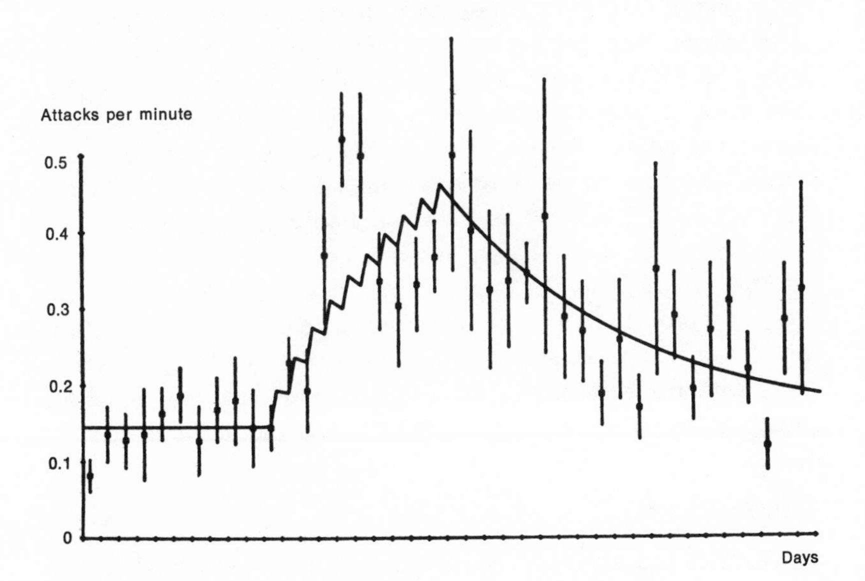

pair spontaneously acquired an appetite for fighting and went out to disturb others." Wickler argues that aggressive behavior must be subject to recall at any time.[59] One can well imagine that the aggressive system might be so constructed, but the observations at our disposal speak against it.

It would seem perfectly reasonable from the constructional point of view to equip territorial animals that often have to defend their territories against their fellows with corresponding motivating mechanisms. The question also arises whether a purely reactive system can be constructed out of ganglion cells that are inherently spontaneous. We know from neurophysiology that nerve cells function spontaneously. Even those that act purely by reflex spontaneously show an increasing readiness to discharge, and many start up spontaneously.[60] In accordance with this, there is a widespread principle of construction by which functional centers in the central nervous system exercise inhibiting influences on each other. Jouvet's study of the paradoxical sleep of cats shows that aggression among mammals is kept in check in a similar manner. Movements of the eyes, ears, whiskers, and paws were observed during paradoxical sleep, and increased electrical activity from certain regions of the brain can be demonstrated. This phase is comparable to the REM phase in the sleep of human beings—the phase, marked by rapid eye movements, in which they generally dream. If a place in the medulla oblongata in cats is destroyed, in 80 percent of cases, the animals show anger during paradoxical sleep.[61] Spontaneous aggression hitherto held in check by other centers obviously breaks through.

We can postulate a drive that produces combat appetence in a number of animal species, but certainly not in all. Also we must expect the drive to be constructed on different principles in different animal species. This does not imply that aggression normally comes to an end only when a centrally stored supply of "energy" is used up. As we have mentioned already, actions are generally brought to an end by special switching-off stimulus situations, and that must also apply to aggression. Fighting generally ends with the withdrawal of one party, and the stimulus of the situation "opponent no longer present" is the end of the matter.

Genetic Factors

Peaceful and aggressive strains of mice have been bred by systematic selection.[62] That this is not a case of the formation and handing down of habits has been shown by experiments in which the young of peaceful mothers were cared for by aggressive mothers, and vice versa. This did not in the least affect the behavior of the young animals; the decisive factor was their genetic origin. The young of the aggressive strain reared by peaceful mothers became aggressive, and the young of the peaceful strain remained peaceful when looked after by aggressive mothers.[63]

4

The "Peaceful" Primate: Territory and Aggression

INTRASPECIFIC AGGRESSION

Our close relative the chimpanzee has repeatedly been called on in recent years to support the theory that man's nature is basically peaceful: chimpanzees living in a state of nature, it has been claimed, have no territories and live in open groups. Reynolds and Reynolds took this view, and so did Jane Goodall in her first publication.[1] Since that time, it has cheerfully been repeated in the secondary literature that chimpanzees do not live in permanent groups and are thoroughly peaceable, and hence that man is by nature a peaceful creature too.

However, as a result of more recent observations of chimpanzees in a state of nature, this view has had to be revised. Goodall later had this to say: "In the early days of my study at Gombe, I formed the impression that chimpanzee society was less structured than actually it is. I thought that, within a given area, the chimpanzees formed a chain of interacting units, with the extent of an individual's interaction with other chimpanzees limited only by the extent of his wanderings. Subsequent observations showed that this was not the case."[2]

From the latest studies, the following picture emerges: chimpanzees live in communities, also known as unit groups, large-sized groups, prebands, and regional populations.[3] Members of these communities know each other and react to outsiders with threats and attack. Within the community subgroups are formed. Since the composition of these subgroups is fluid, this at first created the impression that chimpanzee groups were ill defined and open. (Reynolds and Reynolds state that "chimpanzee groups in the Budongo Forest were not closed social groups. Groups were constantly changing membership, splitting apart, meeting others and joining them, congregating, or dispersing."[4])

This, according to our present knowledge, is true of the members of a community that is itself subdivided. For several years, there is a closer tie between mother and child, and pseudo family ties also seem to exist between certain males and females. And "friendship" seems to exist between certain adults who have more contact with each other than with other members of the community, though there is no clear demarcation.

Communities are sharply marked off from each other and generally live in different territories, which at most overlap at the borders but are often separated by stretches of no-man's-land.[5] The five communities observed by Goodall consisted of between fifteen and forty individuals, occupying living areas with overlapping boundaries. They used these overlapping areas only when their neighbors were absent. Nishida and Kawanaka observed two communities in the Kasakati Depression, about eighty miles south of Gombe. The larger, consisting of about sixty individuals, went once a year to another area, where another group about twenty strong lived. Whenever their neighbors approached, the latter withdrew to the extreme tip of their living area, about twenty miles away, where they remained until their neighbors departed.[6]

Goodall also describes a gradual split that took place in a community of about sixty individuals. Between 1965 and 1967, all the members of the group, which had now divided into two, went to the same feeding place, but from the beginning, some tended to go more to the south and others to the north. The process of splitting may already have begun, or the two groups may have joined up again only at the common eating place. At all events, they ended by splitting up completely when the feeding place was visited less often. At about the same time, a change of alpha males took place, when the old leader, Mike, was replaced. The successor, Humphrey, was afraid of the two highest-ranking males of the southern group and avoided them as far as possible. After the middle of 1972, only one male of the southern group visited the feeding place. Members of different communities avoided each other or chased each other away. Goodall often observed chimpanzees responding to the calls of strangers with countercalls, aggressive display, drumming against trees, or swift and silent evasion. Twice she saw males of one group attacking a strange female.[7] A child of the latter was killed and partly eaten.[8]

On another occasion Goodal saw two females attacking a strange female:

> Just as we were getting some bananas ready for her we noticed Flo and Olly staring fixedly at the stranger, every hair on their bodies bristling.
>
> It was Flo who took the first step forward, and Olly followed. They went quietly and slowly toward the tree, and their victim failed to notice them till they were quite close. Then, with pants and squeaks of fear, she climbed higher in the branches. Flo and Olly stood for a moment looking up before Flo shot up the tree, seized the branch to which the now screaming female was clinging, and, her lips bunched in fury, shook it violently with both hands. Soon the youngster, half shaken, half leaping, scrambled into a neighboring tree with Flo hot on her heels and Olly uttering loud *waa* barks on the ground below. The chase went on until Flo forced the female to the ground, caught up with her, slammed down on her with both fists, and then, stamping her feet and slapping the ground with her hands, chased her victim from the vicinity. Olly, still barking, ran along behind.[9]

Thus chimpanzee communities are closed groups, which does not mean that no exchange of individuals takes place.* Young females in heat often visit other groups, are served by males of that group, and return to their own group when estrus is over. Sometimes they remain with the new group.[10] Chimpanzees differ from other monkeys in this respect. Among baboons and rhesus monkeys, it is the young males that provide for an exchange of genes between groups.

* Closed groups are never so closed that no exchange of genes with individuals of other groups takes place. The point is that bounds and resistances, resulting from the basically hostile attitude to strangers, are set to the exchange of individuals belonging to different groups. That is the difference between closed and open groups, such as shoals of pelagic fish, for instance, which strange fish can join at will.

An interesting observation by Goodall is that chimpanzees generally visit the peripheral areas of their territory only in a larger group. Hearing or seeing strangers triggers off avoidance behavior or aggressive display and fighting. "The chimpanzees gave the impression that they were making deliberate excursions to patrol the boundaries of the community's range," she writes.[11]

Within the chimpanzee community, a distinct social ranking order exists. High-ranking males are identifiable by the fact that they are seldom attacked and that submissive behavior is often directed at them. They have priority in the competition for food, females, and resting places, and make aggressive displays more frequently than others. Rank is secured by fighting and by intimidation behavior. Goodall describes how in 1964 the alpha male Goliath was defeated by Mike, basically by bluff. In threat behavior, chimpanzees make an impression by noise—by striking tree trunks with their hands and feet, for instance. Mike discovered that he could make an especially loud noise by striking empty gas cans he found near Goodall's camp, and he developed this technique to perfection. He ended by pushing three cans ahead of him at the same time, which intimidated Goliath so much that he finally gave up.

Goodall tells us that Mike kept the alpha position he thus attained for six years. At first, he was quite aggressive to other members of the group and attacked low-ranking males often without visible reason, but this intolerance diminished. In 1971, he was suddenly attacked and dethroned by another male known as Humphrey, who was in turn defeated in 1973 after several attacks by the younger Figan, who had the benefit of support from his brother Faben. Without it, in Goodall's opinion, he would have had difficulty in maintaining his alpha position.

Among females, rank generally depends on the presence of adult or near-adult young. Occasionally, two females gang up against a third, but such friendships generally do not last as long as among males.

The repertoire of aggressive behavior is extensive. Some of the common intimidation-behavior patterns are illustrated in figures 8 and 9. The fur on the outer side of the arms and shoulders bristles, which makes the animal appear to increase in size. It strikes the ground or trees and barks and screeches; its long-drawn-out high-pitched *waa* bark is especially impressive. *Hoo* sounds serve to maintain contact between members of the group. Objects are often used in threat behavior. Branches are torn off and dragged along, and sticks, as well as smaller projectiles, are brandished and thrown at conspecifics. The intimidator pursues its victim, and if it catches it, it may strike or bite. It may spring onto its back or trample on it. Smaller victims are sometimes seized and thrown into the air or dragged along the ground. Clashes with members of the group generally do not last longer than half a minute, and it is rare for injury to result.

Males use intimidation displays when they encounter members of another group, hear strangers, or are frustrated, for instance when vainly following a female in heat, or even when climbing a mountain ridge in heavy rain; and finally they use it in struggles for rank.

In her analysis of "The Behaviour of the Chimpanzee," Goodall

Figure 8. A chimpanzee threatening by brandishing a stick. Drawing by H. Kacher from a photograph in van Lawick-Goodall (1971).

Figure 9. Aggressive intimidation behavior among chimpanzees. From van Lawick-Goodall (1975) after originals by David Bygott.

cites twelve typical situations in which aggressive behavior can be observed:

1. In competition for rank, in which a great deal of threatening and relatively little fighting is done.

2. In taking it out on one's juniors: if an animal is attacked by one of higher rank, it often does not dare fight back, but vents its aggression on one of lower rank.

3. If an animal of lower rank does not react correctly to the requirements of a senior: if a female refuses to obey a male, an attack may ensue.

4. If an animal of higher rank does not fall in with the demands of an animal of lower rank, e.g., if a baby chimpanzee wants its mother's breast but is ignored: the young animal may produce an outburst of rage.

5. If their offspring need defense: mothers normally go to their aid. Older siblings also sometimes intervene to help them.

6. If chimpanzees hear strangers: they often resort to intimidation behavior. If they discover strangers in their territory: they attack.

7. If a scrap starts between two animals: others not directly involved often join in. Fighting is infectious.

8. If food or prey is involved: there are sometimes quarrels, but this is infrequent. Squabbles over bananas at the feeding place are more frequent. Goodall sometimes saw fighting about prey.

9. If there is divergent behavior by group members: aggression is released. The assailant's expression suggests that such behavior is felt to be alarming. After a poliomyelitis epidemic, some semiparalyzed animals were able to move only with great difficulty and in a divergent manner.* When a female fell from a tree and broke her

* Pepe, for instance, could only pull himself along on his bottom, dragging an arm behind him. When he made his way to the feeding place for the first time after his illness, the chimpanzees already there put their arms around each other, grinning with anxiety, and stared fixedly at the cripple, who was unaware that he was the occasion for the excitement, but sensed danger. Everyone avoided him. When the far more severely crippled McGregor approached the feeding place, he triggered the intimidation behavior of two males, and finally Goliath attacked him and began striking him on the back. Had Hugo van Lawick not intervened to protect him, another male would have attacked him too. Later, the chimpanzees got used to the sight of the cripple, but they excluded him from all social activities. Goodall described this as follows:

> There was one afternoon that without doubt was from my point of view the most painful of the whole ten days. A group of eight chimps had

neck, the other chimpanzees were frightened and began making aggressive displays around the dead body, and a male threw a rock at it.

10. If a chimpanzee is in pain: it becomes irritable; a male with a broken toe attacked young chimpanzees that played in his vicinity.

11. If two males are courting an especially attractive female when she is in heat: quarreling occasionally breaks out. This leads to the temporary formation of "consort pairs," an important recent discovery.

12. If chimpanzees are confronted by baboons and human beings: they use intimidation behavior in basically the same fashion as against conspecifics. (Kortlandt[13] and Albrecht and Dunnett have observed that chimpanzees threaten stuffed leopards in the same way.[14])

Aggression unquestionably plays a big part in the life of chimpanzees, though within the group it is generally limited to clashes at the level of intimidation behavior. There is a positive correlation between intimidation behavior and high rank.[15] Charging displays are confined to the males. Van Hooff showed by cluster analysis that there is an attack system and a bluff system, which are closely interrelated but nevertheless represent distinct subsystems.[16]

Chimpanzees are certainly potentially very aggressive, as is shown by the large number of their appeasement behavior patterns, which incidentally strike one as very human. Lower-ranking animals, for instance, stretch out their hands when they want to appease an animal of higher rank. If the latter touches the outstretched hand,

---

gathered and were grooming each other in a tree about sixty yards from where McGregor lay in his nest. The sick male stared toward them, occasionally giving slight grunts. Mutual grooming normally takes up a good deal of a chimpanzee's time, and the old male had been drastically starved of this important social contact since his illness.

Finally he dragged himself from his nest, lowered himself to the ground, and in short stages began the long journey to join the others. When at last he reached the tree he rested briefly in the shade; then, making the final effort, he pulled himself up until he was close to two of the grooming males. With a loud grunt of pleasure he reached a hand toward them in greeting—but even before he made contact they both had swung quickly away and without a backward glance started grooming on the far side of the tree. For a full two minutes old Gregor sat motionless, staring after them. And then he laboriously lowered himself to the ground.[12]

the former is reassured. A female who has just given birth takes her baby to the group, showing it to each member in turn with an expression of anxiety. The mother stretches out her hand to each member of the group in turn, and as each touches it, the mother's tension disappears, and she goes on to the next. As a stranger to the group, the baby could easily release aggression; hence it is introduced appeasingly. Other actions that have an appeasing effect are crouching in front of those of higher rank, kissing them, and sexually presenting the anal region. Higher-ranking animals respond to such submissive behavior by briefly feeling the other party, touching the outstretched hand, kissing or embracing the junior animal, or briefly delousing it. These behaviors reassure the junior animal and create a relaxed atmosphere. The frequency with which appeasing behavior patterns can be observed is further evidence of the great aggressive potential of chimpanzees.

This is certainly not the result of chance, but of selective pressures. We have already discussed the advantages resulting from the demarcation of groups, their territorial spacing, and the formation of a ranking order with a leader and reproductive success based on rank.

Gorillas live in small groups of up to thirty individuals ruled over by a senior male. They may consist of polygynous families. The senior males are more than twelve years old and are distinguished by silvery-gray back fur. Females outnumber males in a group. In contrast to the practice among chimpanzees, among gorillas it is the young males that leave the group and seek entry to other groups. Members of different groups threaten each other and keep out of each other's way. Struggles for rank often take the bloodless form of the animal's beating its breast with its fists. Staring is one form of threat behavior, and submission is indicated by looking away, dropping the head, and crouching down.[17]

Too little is known about the group structure of the orangutan in a state of nature to say more about it here. Gibbons are known to defend their family territories by calls, threats, and fighting. Carpenter reports that many gibbons show scars resulting from fights with conspecifics.[18]

The foregoing makes it evident that the frequent statements in the secondary literature about the special peacefulness and nonterritoriality of the primates that are closest to us—chimpanzees in particular—are misconceived. An example of the uncritical repetition of old, long superseded information is provided by Schmidbauer, who goes on peddling the story of friendly, peaceful primates living in open groups (he did so as recently as 1974): "Chimpanzees and orangutans do not live in permanent groups," he writes. "They form temporary aggregates in which social ties are friendly and are based on similarity of sex or age, sexual attraction, mother-child relationships and possibly sibling relationships (V. and F. Reynolds, 1965). One of the surprising characteristics of chimpanzee society is that—even when the temporary group dissolves—relations are marked by affectionate greetings and meetings as soon as individuals meet again."[19]

Pretty well everything in this passage is incorrect, and the conclusions drawn from it are accordingly unsound. These are: "I postulate here that this characteristic of open in contrast to closed groups has characterized the mainly pongid-hominoid line since the premonkey stage of the Eocene and is responsible for the form that human society has assumed."[20]

PREDATORY AGGRESSION

Chimpanzees hunt other mammals, for instance, small antelopes, colobus monkeys, and baboons. They either seize the quarry and smash the skull against a rock or tree trunk or kill it by biting the head and back of the neck. Males sometimes hunt together and act cooperatively. Goodall describes a male climbing a tree after a colobus that had become separated from its group while his companion waited below to cut off its retreat.

Meat eating was certainly discovered relatively recently by chimpanzees. Hunting behavior is unknown among other primates,* and both in their physical structure and in their behavior, chimpanzees lack specific adaptations to hunting. That does not mean that it has little importance in their lives. They hunt larger mammals systematically and obviously deliberately. Between 1960 and 1970, ninety-five successful and thirty-seven unsuccessful hunts were observed in the Gombe territory, and it was possible to identify the quarry in fifty-six cases. The table below shows that monkeys accounted for 65 percent of it.

* Except among baboons, which do not hunt systematically, but will take advantage of a chance opportunity to kill a young gazelle or a hare.

PREY CAUGHT BY CHIMPANZEES IN THE GOMBE TERRITORY BETWEEN JUNE 1960 AND AUGUST 1970

Prey	*Number*	*Percentage*
Baboons	21	38
Red colobus monkeys	14	25
Guenons		
Cercopithecus ascanius	1	1
Cercopithecus mitis	1	1
Bush pigs	10	19
Bushbucks	9	16

There must certainly also have been other hunts that escaped observation. From his studies, Teleki concludes that in their role as predators, chimpanzees play a part in the economy of the area. The dietary role of meat eating has not been investigated. Quantitatively it is negligible, for it provides only 1 percent of the total, but qualitatively it provides high-grade protein. At all events, its social importance is great. When the hunt is over, quarreling may break out over the prey, which is sometimes torn into several large pieces in the process. But if an animal maintains possession of his prize for a short time, the others respect this. They sit around him and beg, and he shares it out. High-ranking animals cleverly hand out small portions, thus prolonging their possession of the prize and remaining the center of attention. Most of the members of the group who are present get a share. In one instance reported by Teleki, the senior male, Mike, between 8:00 a.m. and 7:30 p.m. distributed portions to thirteen of the sixteen adults and adolescents present.[21] In this way, animals of high rank are in a very positive sense the group's courted center of attention (figure 10). We emphasize the great social importance of hunting and sharing out prey. In many hunting and food-gathering tribes, the group bond is strengthened by this sharing out.

Chimpanzees show a striking preference for the brain of the quarry. They generally open the cranium by biting it from above and extract the contents with their fingers. They often end by wiping the cranium clean with chewed leaves. This delicacy is never shared with another animal.

Two recently reported cases of cannibalism are worthy of note. In both, adult males captured a young animal and began eating it while it was still alive. In the second instance, Bygott watched the whole process from beginning to end. A group of Gombe chimpanzees came across two females. Five adult males immediately attacked the older female. Bygott was sure that he had never seen this female in the many hundreds of hours he had spent observing chimpanzees in the area, and he assumes that she was a stranger to the assailants too. The struggling group disappeared into the bush, and when the animals reappeared, the female and the dominant male were missing. Humphrey (one of the chimpanzees also observed by Goodall) was holding a struggling young chimpanzee about eighteen months old by the leg; this animal which was bleeding from the nose, was also unknown to the observer. Humphrey repeatedly struck its head against a branch, and after three minutes started biting flesh out of its thighs. Then Mike, the oldest male, approached and tore out one of the animal's legs, which he spent the next hour and a half eating. When Humphrey stopped eating, the body was intact except for parts of the leg. Humphrey began grooming and delousing it. After this exploratory play, he became aggressive; he bit the body and flung it to the ground. Eventually he went away, leaving the body where it was. Mike then took it over, spent about an hour and a half chewing it, and gave it back to Humphrey when he returned. Humphrey flung it against a rock and let it lie there. Another male, Romeo, spent ten minutes gnawing at it. Then Humphrey and Mike returned, picked it up, dropped it again, and left it to the males Figan and Satan, who played with it and also deloused it. When they finally left the body after six hours, only the legs, one hand, and the genital region had been eaten, though the body had passed through six chimpanzees. "This suggests," Bygott writes, "that the prey was less attractive as a food object, and more an object of curiosity, than other animals which Gombe chimps have been seen to eat."[22] In fact, the behavior of the chimpanzees toward their dead conspecific was rather ambivalent, as the alternation between aggression and social delousing shows.

Predatory aggression against a conspecific is certainly unusual among predators, and the rarity of reports of cannibalism among chimpanzees shows that it is not normal either. For an event such as the killing of a strange young animal to have occurred, there must have been a combination of special circumstances, such as a surprise meeting with a strange group coinciding with a high degree of emotional excitement. Once an animal is dead, it ceases to present inhibi-

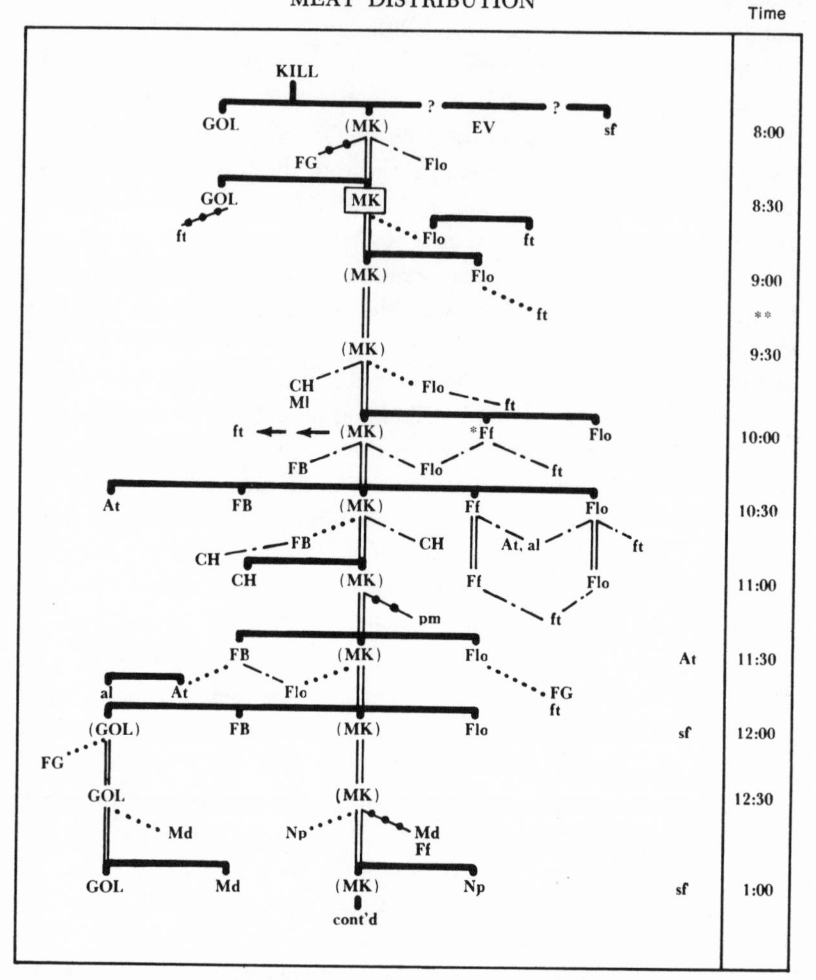

* Females in estrus-Ff, with almost complete tumescence

** Chimpanzees returning to the camp for 30 minutes

*** FG stealing meat from GOL in surprise attacks

Figure 10. Charts showing the distribution of meat by a senior male chimpanzee. From Teleki (1973). Each chimpanzee is identified by an abbreviated form of its name; e.g., MK is Mike, the senior male chimpanzee.

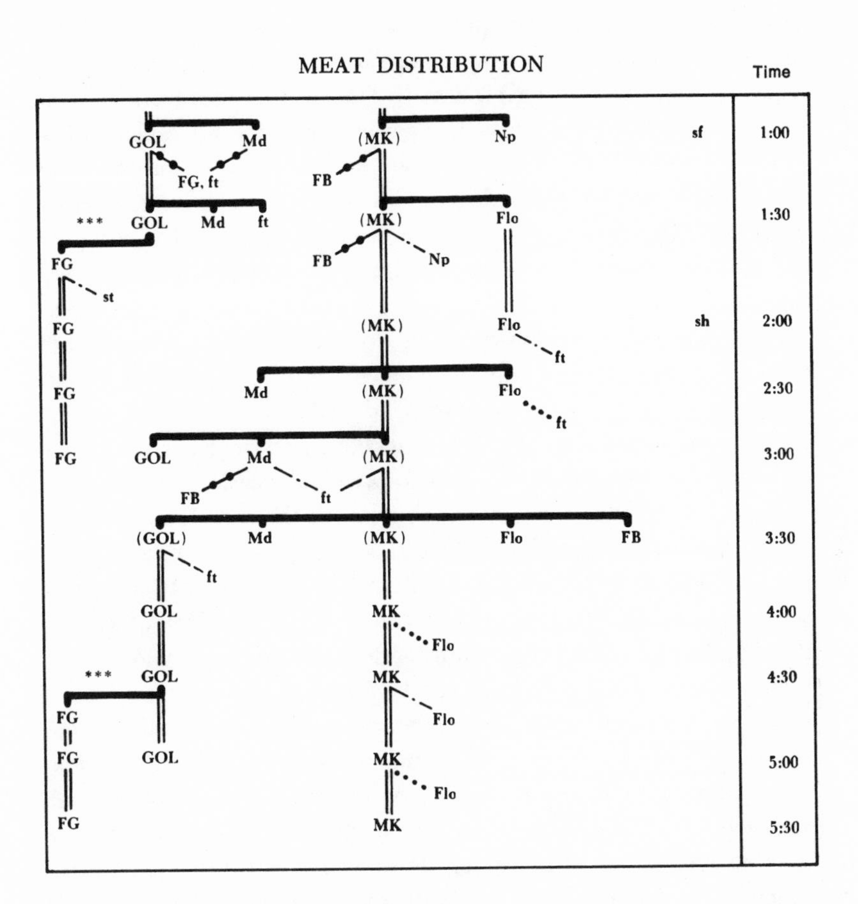

Division at the beginning and later of major portions

( ) Possessor of major portions

Consumption of the brain

Meat touched but not taken

Approach to and observing of the possessor of meat

Unsuccessful begging for meat by gesture and sound

Successful begging by gesture and sound

Brief chewing of meat belonging to another

Meat freely offered or given to another

tory signals; it becomes just prey. The instances described show that the boundaries between prey and conspecifics are not so sharply drawn as they are among predators. This also applies to their relations to prey of other species. Goodall reports that adolescent chimpanzees play with adolescent baboons, and it has also been observed that hunting chimpanzees do not always stalk their quarry noiselessly, but often threaten it with aggressive display, which is certainly not appropriate to the purpose. Judging by its motivation, the behavior of chimpanzees to their prey probably derives from intraspecific aggression. Behavior patterns of intraspecific aggression are used in capturing prey, and interspecific aggressive behavior patterns—such as striking the quarry's head against tree trunks or stones, threatening with sticks, biting the backbone and the head, and stone throwing—are sometimes used in intraspecific clashes though, and this must be emphasized, this is by no means the rule. It seems more likely that these cases are accidents deriving from a separation of the two fields that is phylogenetically not yet complete. In man, this separation similarly seems to be not very sharp. The lack of a clear dividing line between prey-capturing behavior and aggression is also pointed out by Kortlandt; however, he certainly goes too far when he says that chimpanzees kill, not for the sake of the prey, but to demonstrate their ability to kill to other members of the group—intimidation by terror, as it were.[23] Teleki's observations do not permit such a conclusion. The cooperation demonstrated by chimpanzees in stalking and killing shows that the objective is the prey. A social function is evident only in the dividing up of the spoil that follows.[24]

AGGRESSION AGAINST PREDATORS

To the best of my knowledge, no one has yet had the good fortune to observe chimpanzees faced with predatory enemies, but Kortlandt's experiments with chimpanzees in a state of nature have made it seem probable that they attack them, using weapons to do so. If a stuffed leopard is placed in an area inhabited by chimpanzees, they approach it, threaten it, aim lumps of earth and stones at it, and attack it with wooden sticks; they approach it upright and aim well-directed blows at it.

In the course of these experiments, Kortlandt noted remarkable differences in the behavior of savanna and forest chimpanzees. Chimpanzees of the open savanna throw missiles with a good degree of accuracy and strike well-aimed blows with sticks, while forest chimpanzees, though they show the same basic behavior patterns, throw badly or aimlessly and are not very skilled in the use of sticks. These marked differences indicate that the use of sticks and missiles as weapons developed in the relatively treeless savanna; in the forest, these weapons would often be obstructed by branches.[25] Kortlandt and Kooij assume that the forest chimpanzees used to be more skillful in the use of these weapons, but gradually lost their skill after being forced into the forest. They suggest that the ancestors of present-day chimpanzees in general developed much further in the human direction, but were defeated in the competition for the savanna habitat with man's prehuman ancestors and were forced into the forest areas, which do not promote the evolutionary transformation of primate into man that we call hominization.[26]

USE OF WEAPONS

Chimpanzees use objects as weapons in contests with conspecifics and others. In clashes with their predatory enemies, sticks and missiles are chiefly used. No one has yet seen a chimpanzee hitting or hurting another with a stick, and no one has seen them hunting with sticks. Unlike men, who use weapons to kill their fellow men, chimpanzees use sticks primarily against predators. In intraspecific relations, they use them to threaten, brandishing them as if to strike.

Kortlandt has described the various forms of agonistic intimidation behavior and fighting with weapons. In the course of his remarkable survey, he also discusses the distribution of behavior patterns among chimpanzees, gorillas, orangutans, and man. Here is a brief summary:

THREAT BEHAVIOR USING STATIONARY OBJECTS

1. Striking the ground or branches of trees with the flat of the hand. Practiced by all four species.
2. Stamping the ground or striking a tree with the sole of the foot. Known among chimpanzees, gorillas, and men.
3. Rapid drumming against resonant objects, either with both hands or with both feet, using hands and feet alternately. Occurs among African apes and man.
4. Rapid shaking of branches with the hands or all four limbs. Known among gorillas, chimpanzees, orangutans, and some lower

monkeys. According to Kortlandt, man frequently shakes an opponent in this manner.

5. When the animal is up a tree, swinging branches slowly in the direction of the enemy, either standing or on all fours.

6. Vigorous tearing up of plants or tearing down of branches while the animal moves intimidatingly over the ground or through the branches. Known among African primates and in altered form among man, the roots of whose destructiveness may perhaps be found here.

THREAT BEHAVIOR USING MOVABLE OBJECTS

1. Brandishing a branch or stick. Observed only among African primates and in man. Savanna chimpanzees adopt the upright stance for this, while forest chimpanzees move on three legs, even outside the forest. Derives from the behavior described in example 6 above, but in some cases is to be regarded as the threat to strike (see 1 and 4 in the next section).

2. From the above, Kortlandt distinguishes the stick shaking carried out with a rapid, rhythmical up-and-down movement that perhaps derives from example 4 above. Is carried out on the ground by the chimpanzee.

FIGHTING WITH WEAPONS

1. Breaking off and throwing down of branches and other objects. Carried out with an overarm movement, either standing (on two or three legs) or from a tree. The flat of the hand is held forward, and the aim is generally bad. Observed among chimpanzees, gorillas, and orangutans. In man, ritualized in childhood as the breaking of objects for the purpose of intimidation.

2. Underarm throwing, i.e., with a sideways or upward movement of the arm, thumb pointing forward, arm slightly bent. Carried out standing or moving on three limbs. The missile is aimed, and its curved trajectory is taken into account. Kortlandt states that when a chimpanzee throws something at a conspecific, it aims slightly to one side, that is, to miss. I find this very remarkable. This has also been noted when Bushman children throw things at each other. This type of throwing has been observed only among African primates and man. In man, it is sometimes known as 'girlish' throwing. It is much more developed among savanna than among forest chimpanzees. Kortlandt conjectures that the original biological function may have been throwing sand in an enemy's eyes and that throwing pieces of rock may have been a secondary development from this.

3. Overarm throwing. Differs from the first type in this section in that the hand does not travel high over the shoulders; the arm at first is bent, but stretches with the throwing motion, the little finger is pointed forward. The throw is accompanied by a twisting motion

of the spine, hips, and legs, and there can be great power behind it. This is a typical specialization in man.

4. Striking with a stick, which is raised and brought down on the opponent (see the experiment with the leopard, p. 74). The force behind such a blow delivered with a weapon about two yards long at a speed of between fifty and fifty-five miles an hour is considerable. Well developed among savanna chimpanzees, but only rudimentary among their forest counterparts. One-year-old humans strike playfully in this manner, using the 'play face' (p. 91).

5. Underarm stabbing, i.e., stabbing in an upward direction, with thumb pointing forward. So far observed only as explorative behavior and not in an agonistic context. Kortlandt points out that there are strong inhibitions in man against using this movement in fighting with conspecifics; in military and jujitsu training, these inhibitions have to be overcome by practice with rubber daggers and straw dummies.

6. Overarm stabbing, with little finger in front. As a fighting technique, observed only in man. Fewer inhibitions are associated with it. Frequently used by humans in attacks of rage against their conspecifics. Kortlandt explains the difference between overarm and underarm stabbing by suggesting that the downward stabbing motion with the blunt hand-ax of the Paleolithic era was less dangerous than an upward stab. " 'Thou mayest stab, but thou shalt not kill' was apparently the sixth commandment for palaeolithic man," Kortlandt observes.[27] Schoolboys use this movement when they go for each other with their fists.

5

Aggression in Man

As we have seen in the preceding chapter, our closest relatives, the great apes, have considerable aggressive potential and are also territorial. Until a few years ago, it was believed that chimpanzees were an exception to this, but recent studies have shown otherwise. This strongly suggests that our human aggressivity may be an ancient primate heritage. Many have expressed this view, but until a short time ago, it was vigorously denied. Extreme adherents of the environmentalist theory maintain that human aggression is solely a result of social conditions and that the only constitutional factor is the disposition to orient oneself in accordance with a social model and to react aggressively to experiences of deprivation (frustrations). They also assert that children learn that aggression leads to success and, on the basis of this experience, use it instrumentally when they want to have their way. There is no more to it than that, they claim, and above all, they deny that the primary motivation is a drive.

Now, I do not have the slightest doubt that social conditions have great significance in the formation of man and of his attitude to aggression in particular, but I regard as inadequate all theories that allot only a minor role to the determining factor of heredity. I propose to substantiate my view in the course of this book. Reading the relevant literature gives me the impression that the rejection of biological determinants is often based on the fear that acknowledging them would open the way to fatalism—a conclusion I am unable to follow (see pp. 165 ff).

The spectrum of human aggressive actions is a broad one. Man can release his aggression directly against his fellows by striking or insulting or mocking them. He can do so indirectly by speaking ill of them or setting traps for them. He can act aggressively by refusing social contact: by refusing to talk or to help them. Aggression can be directed against individuals or groups, either in ideological conflict or in war. What is common to all these actions is that, with their aid, pressure is exercised on a fellow man or group that leads ultimately to its departure from the scene or to its submission to a person or group of high rank or to the group norm. Aggression often causes pain, and in man it can be destructive.

INTRAGROUP AGGRESSION: ITS FORMS AND EFFECTS

Aggressive clashes between members of a group are by no means uncommon, but generally they are not destructive. There are a number of typical situations in which intragroup aggression appears. In enumerating them here, we shall cite examples from our own cultural area. Later, we shall see that basically the same patterns also appear in other cultures.

Humans quickly develop spatial habits. This is readily evident when a family sits down for a meal. The places around the table occupied by different members of the family are very conservatively adhered to; by tacit consent, everyone respects everyone else's place. These spatial habits are clearly developed by the age of two; children of that age are obviously upset if they have to change their accustomed place. In certain circumstances, such territorial rights are established very quickly indeed, for instance, in a railway compartment, where the firstcomer's claim is generally recognized—so much so that subsequent arrivals politely and appeasingly inquire whether a place is still free, even though this is obviously the case. Before taking a seat at a restaurant table at which others are already seated, one asks their permission, and approaches their table only if no other tables are free. To act otherwise would in many cases be felt to be obtrusive, unless all the guests were members of the same party, for instance.

The distance that one individual keeps from others depends on the strength of the sense stimuli that come from them. For experimental purposes, Nesbitt and Steven placed strikingly dressed men and women among individuals in a line for tickets to an amusement park. Those behind them in the line kept a greater distance from them than from conservatively dressed persons in the same line. Perfume and after-shave lotion had the same spacing effect.[1] The fact that in modern mass society men tend to dress less strikingly than in earlier periods can be regarded as an adaptation to crowded cohabitation.[2] The display coloration of many birds and fishes has a similar distancing effect.

Felipe and Sommer examined individual distancing in libraries. They sat at occupied tables even if other tables were free and, as if by chance, in close proximity to the readers already there. The latter invariably tried to withdraw from the intruder and, if this was impossible, erected barriers of books, rulers, and other objects against him. If these maneuvers failed, the reader would leave the table, often with manifest signs of displeasure.[3] The distances individuals keep between themselves and their fellows varies culturally, but there is a universally valid rule that physical contact is permitted only in precisely defined situations and not at all times.

Esser and Palluck observed the behavior of twenty-one mentally deficient boys in an amply compartmentalized experimental room in which each boy had a space of his own. At first they quarreled, but when the room was finally divided up between them, they behaved very peacefully; only a minimal amount of threatening was necessary to make sure of keeping a place. The territorial behavior of severely handicapped children (with IQs under 50) was more marked than that of normal children, and it was hardly influenceable by the verbal punishment that was effective in other cases.[4] The dividing up of space results in a social order that gives the children a sense of security. Everyone knows his place and knows he will be left in peace in it, or that he will be able easily to hold his own against intruders, for even among these mentally deficient children, there is a presumably innate acceptance of the principle that the first occupier of a place enjoys a priority that must be respected.

Edney found in Connecticut that persons who displayed notices on their property such as Private Property, Keep Out, No Trespassing, Warning: Keep Out had lived, or proposed to live, longer in the area than their neighbors who displayed no such notices. Also, when inquirers called on them, nominally to ask them questions about pollution of the environment, they answered the bell more quickly than their neighbors, thus showing greater sensitivity to the presence of a stranger on their territory.[5] The same author also examined the distribution of groups on a homogeneous stretch of beach in Connecticut, and here too territorial grouping was evident.[6]

Such territorial behavior within a group establishes a certain order and stability and contributes to the individual's sense of well-being. As people are thrown together in modern working and living conditions, stable and dependable relations between them are advantageous. These are promoted by lasting territorial ties.

Children often quarrel over the possession of a coveted toy. At the crawling stage, they usually try unceremoniously to grab it from its possessor, who generally defends himself by clinging to the object and protestingly appeals for the aid of third parties. He may try to flee with the object still in his possession. Those robbed of their possession protest (weep) and often go over to the counterattack, trying to grab the object back or directly assaulting the evildoer, striking or scratching or pulling his hair and very often pushing him over. These strategies are adopted spontaneously and are not confined to the children of our cultural area. Presumably, we have here an innate behavior pattern in which the technique of jostling and pushing over is especially noteworthy, as it seems to be a specifically human pattern of behavior that could have developed only with the upright stance.

Older children respect possession. If a child takes another child's toy, the latter may manage to get it back by verbal protest and repeated insistence that the toy is his. Generally a child who wants an object that is in the possession of another will ask to be lent it; only after the request has been refused will he grab it, or often actually attack the possessor. In other cultures also, refusal to share causes offense; it is a breach of good manners that triggers aggression. We have mentioned that chimpanzees that have managed to keep freshly caught prey for some minutes then "possess" it, i.e., no longer have to defend it, because the others obviously respect its property rights and beg for shares instead of grabbing or stealing them. It is worth noting that a functionally comparable social inhibition against attacking another because of his "property rights" has been shown to exist in the sacred baboon (*Papio hamadryas*), among whom males respect the possession of each other's females. It may be that corresponding social inhibitions exist among beings that prevent perpetual strife for the possession of objects as well as of partners and individual territories. The advantages of such inhibitions from the point of view of selection are immediately obvious.

The Influence of Strife and Cooperation

In general, rivalry, whether for reward or for other objectives, leads to aggressive demarcation and hostility between two competing groups. Sherif and Sherif studied the development of relations between groups of eleven- and twelve-year-old boys in holiday camps. The experimenters acted as camp leaders. At first, the boys, who were previously unknown to each other, slept in the same wooden hut and were free to make their own friends. After two or three days, small groups of from two to four boys had formed, and when a boy was asked to name his best friend, he was able to do so. They were then split up into two groups, in which friends were deliberately separated. The original friendships ended, and new preferences were formed within the group. The members of each group had to cooperate in such jobs as putting up tents, carrying canoes, cooking, and other things. Group structures on the leader-and-led pattern developed independently in each group. Special individual skills came into play and it was notable that the two groups developed different styles in tackling tasks. This was especially evident in later experiments that led to conflict between the groups. The group that took a tough line enjoyed physical clashes, while the other resorted to the aid of prayers to bring about victory. The tendency to cultural pseudospeciation showed itself in the development of group-specific jargon, secrets, and jokes.

As soon as the distinct character of each group was sufficiently established, competitive situations were brought about. The two groups played baseball and engaged in tugs-of-war and other competitive games, and the winners were rewarded with much coveted prizes. The experimenters assumed that if two groups competed for an objective that could be won only at the loser's expense, hostile tensions would develop; this was confirmed. The losers taunted the winners and were answered in kind, and finally "revenge" expeditions were organized in which the parties bombarded each other with green apples. At the same time, the members of each group developed a negative opinion of the members of the other. They no longer liked them, and that made them feel more closely bound to their own group. Each side overrated its own achievements and underrated those of the other.

After a competition to see which group could pick the most beans, the boys were shown photographs allegedly picturing the beans that had been picked. The number of beans was always the same, but each group was shown first what was stated to be its own collection and then what was stated to be the other's. The boys underestimated their competitors' score and overestimated their own.

The competition-induced polarization of the two groups was finally broken down by joint operations in which both were given an overriding objective. The organizers staged a fault in the camp water supply, and both groups took part in tracking it down. After it had been repaired, the two groups relapsed into their negative attitude to each other. The organizers next announced that they did not have enough money to pay for a very popular film that everyone wanted to see. The two groups discussed the situation together, the necessary funds were raised, and both watched the film together. By means of these and similar situations, the old hostility was gradually broken down. Before these last experiments were made, the two groups were temporarily united by being given a common enemy, but this strategy was not pursued further.[8]

There are striking structural resemblances between the behavior of rival groups of children and those of rival nations.

Everyone knows that human beings compete for the favor of other human beings. When young men compete for a girl's favor, they sometimes come to blows. When a bond such as that between a married couple has been established, it is defended. Jealousy is a universal emotion shared by both sexes; it leads to strife between women in polygynous families, even though in the cultures in which polygamy occurs, it is mitigated by the fact that women have less fear of losing their men. Strife nevertheless occurs, and in many polygynous cultures, men like to marry sisters, as the sibling bond is said to diminish aggression.

Members of a group often compete for the favor of persons of higher rank. Finally, we know that children compete for their parents' love. This sibling rivalry is especially marked in young children after the birth of another child. This is another phenomenon observable in many cultures.

When a child is attacked by an adult, other adults almost automatically go to its aid. I once saw a man chasing a young thief, aged about ten. The passers-by immediately took the boy's side. A basic protective reaction must be at work here, comparable to that displayed by many vertebrates in defense of their brood. Parents defend their children, and interestingly, even infants defend their parents. There is an unconsidered, almost reflex, quality about this spontaneous intervention on behalf of the individuals closest to us that points to a response to innate patterns (p. 48). "Family [or 'group'] in danger" is accordingly a cliché much used when it is desired to rouse the combativeness of members of a group.

The Importance of Rank and Status

Human beings aspire to recognition by their fellows. They try to distinguish themselves by special achievements and thus at least temporarily to be center of the group's attention. This aspiration is shared by the representatives of egalitarian societies, though in such societies, the individual's efforts do not lead to the formation of a fixed hierarchy with an absolute claim to leadership. The good hunter is respected for his hunting skill, and is consulted when his advice is needed. The same applies to the good trance dancer, storyteller or maker of ostrich-eggshell necklaces.

Classifying people according to their gifts is certainly adaptive. Being respected means securing the attention of one's fellows, and that is important to the individual. Chance was to the best of my knowledge the first to point out that among the higher apes, an animal's rank can be most quickly discovered by reference to the "attention structure," and the same applies to us human beings.[9] One only has to look at the seating order around a table. The highest in rank generally sit where they can be seen by everyone else: at the narrow end of a rectangular table. Aspiration for rank involves competition, but superiority is generally not attained by the use of force. We have mentioned that among chimpanzees there is a correlation between seniority and intimidation display, not between seniority and physical aggressivity. This also applies to other nonhuman primates, and it surely applies to men too. In the most varied cultures, individuals attain and keep high rank by skill in the distribution of goods, for instance. Other essential qualifications for high rating in the ranking order are the ability to keep the peace and the ability to gain friends, both of which are positive social characteristics. When high rank is the result of aggression, ritualized forms of intimidation are generally used. Men even use their material possessions for this purpose. Ostentation and extravagance are characteristics not only of the Kwakiutl potlatch* or receptions at princely courts, but also of state receptions in Western democracies. Anthropologists talk very appropriately of prestige economies in which resources are squandered for the sole purpose of impressing rank. Sometimes this assumes the character of a ritualized fight, particularly when different allied groups engage in rank displays.

In hierarchically organized societies such as our own, status symbols have developed by which people's rank can be recognized, as Vance Packard observed. Such status symbols include office furnishings, jewelry, cars, and clothes. The things do not have to be good so long as they are expensive. Those who cannot keep pace are not "in." It does not say much for these "money elites" that they attach value to such banal superficialities. But their striking behavior attracts attention, and they have their public and their imitators, as the newspapers show.[10] Morris has pointed out the remarkable phenomenon of rank mimicry. The ways of the nobility and so-called high society and the clothes they wear are copied by fashion and imitated by the masses, which forces those they imitate to preserve their exclusivity by thinking up something new.[11]

The aspiration for rank sometimes leads to pathological degeneration, as when someone makes himself the center of public attention by committing a crime. A more innocuous phenomenon is the creation of surrogate pyramids for themselves by people who have no hope of attaining any kind of high social rank: they climb to the top of these pyramids, whether as pigeon fanciers or collectors of matchbox tops.[12] Such observations point to a strong motivation underlying the aspiration for rank.

Leaving aside such marginal phenomena, it is clear that the group benefits. Basically the outcome is selection by merit, resulting, at any rate in primitive societies, in the production of more offspring by those of higher rank. The latter also perform a number of functions in the group; one we have mentioned is keeping the peace. Also, thanks to the authority they enjoy, they play a leading role in decision making.

* Feasts given by the Kwakiutl Indians at which they smashed valuable copper plates, poured expensive oil on the fire, smashed boats, and killed slaves in order to impress their guests. When their hospitality was returned, their hosts had to behave just as extravagantly to save face.

As I have pointed out elsewhere, every ranking order assumes not only an aspiration for rank but also a tendency to subordination and obedience. The latter is not just the result of enforcement by strict upbringing. Stayton, Hogan, and Ainsworth found that children in a friendly, accommodating environment are ready to obey without the application of any pressure:

> These findings cannot be predicted from models of socialisation which assume that special intervention is necessary to modify otherwise asocial tendencies of children. Clearly, these findings require a theory that assumes that an infant is initially inclined to be social and [somewhat later] ready to obey those persons who are most significant in his social environment. Such an alternative viewpoint is offered by the ethological-evolutionary model of early social development presented by Bowlby and Ainsworth.[13]

In man the tendency to obedience is very marked, and blind obedience, as was demonstrated by Milgram's experiments, is a very dubious phenomenon.[14] The aberrations of the aspiration for rank—such as addiction to luxury goods and other forms of self-display—must also be regarded as dubious. When we interpret extravagance as magnanimity and generosity and are impressed by display, we are obviously reacting to very ancient stereotypes. At one time, there was certainly a connection between an individual's ostentatious appearance and his personal qualities, but that is no longer the case today.

There are regulatory mechanisms that tend to balance things out. Envy, according to Schoeck's remarkable study, works against the externals of intimidation behavior and has a compensatory effect, chiefly in smaller, more individualized societies, like that of the Bushmen, for instance. An individual who has something is forced to share it by the envy of others, and it is impossible for an individual to raise himself very far above the rest by virtue of his possessions.[15] In our own society, this factor, and the redistribution of property enforced by it, exercises a similar balancing function, and it has certainly been a decisive factor in all social revolutions. Finally, the aspirations of the lower ranks guarantee that hierarchies do not get fossilized and that the ranking order remains dynamic.

The more anonymous a society becomes, the more problematic become the criteria by which the top-rank "leaders" are selected and the means by which they maintain their position. Since no one knows political candidates personally, one has to accept what they say in their speeches or what the press says about them, and this, as we know from history, is not always correct. An interesting feature of the selection process is that consent is actually sought; this obviously forms part of the behavior for which we are programmed in advance. We show our consent and are then willing to follow, but without calling ourselves to account for it.

I suspect that domination based essentially on violence is not in accordance with man's nature and leads to counterviolence. Certainly there have been tyrants who have ruled with the aid of bloody terror, and perhaps emergency situations are conceivable in which this extreme behavior may be advantageous (see my observations on bonding through fear in *Love and Hate*). But I have the impression that no domination can maintain itself through terror in the long run. At all events, I know no historical study that shows that it has.

Ranking orders also develop among groups of children, as Hold has noted (see p. 46). High-ranking children are the focus of a group's attention: they are the leaders at play and act as peacemakers.[16]

Exploratory Aggression and Educational Aggression

Aggressive behavior often serves the purpose of testing out the boundaries of the social field of action. Children direct their aggression to finding out how far they can go; from the response, they discover the limits of tolerance and what their culture expects of them as the standard of behavior. In the absence of response, exploratory aggression tends to escalate.[17]

Exploratory aggression also plays a part in competition for rank in small groups. The weaknesses of other parties are discovered by this means. As the student unrest in recent years demonstrated, escalation can easily take place if an establishment that has grown unsure of itself is incapable of constructive resistance. The same applies to the conflict of generations, the tug-of-war between the forces of cultural change and those that resist it (p. 25).

When aggressive social exploration is at work, provocation, according to Hassenstein, cannot be bought off merely by granting the demands that are made. On the contrary, surrender leads only to further escalation. The aggressor is trying to establish his inferiority or superiority by a trial of strength, and a voluntary self-limitation without a fight is not to be expected. Hassenstein sees in this fact one of the reasons that an overcompliant upbringing on the one hand, and a concessionary school or university policy on the other, lead, not to satisfaction, but to an increase in aggressivity all around. He continues:

> This is not a situation that will change in the future. For what the strategy of surrender, of which smoking rooms in schools and youth hostels—the very opposite of a responsible health policy—can stand as a symbol, generally comes up against is not aggressivity determined by frustration, but socially exploratory provocation. This cannot be appeased by surrender, for after their demands have been met, the young persons concerned are left without landmarks in an uncertain, unstable situation out of which they try to break by new activities.[18]

It should not be forgotten, however, that this kind of aggression is also an important positive force in cultural development, since by means of it, changes are brought about against resistance.

Offenses against the rules of social life involve punishment. In children's play groups, the older children correct those who break the rules of the game: they scold them and often punish them physically. The same applies to the social life of adults: lawbreakers are pursued and punished.

The type of aggression described by Trivers as moralistic also has an educational function. Moralistic aggression developed to protect the group against "cheats" who gain personal advantage from an altruistic relationship. Injustice, unfair behavior, and meanness release this form of aggression. According to Trivers, it is certainly advantageous from the point of view of selection that human beings should react violently against all forms of cheating.[19]

Educational aggression is often aimed at fitting the pupil into a low rank. This applies both to military drill and to the initiation rites practiced by many peoples. In these, young people are often humiliated and tormented in surprising fashion and are instructed only afterward. Finally they are accepted as full members of the community. Underlying these initiation rites, there must be an awareness of the natural willingness to learn from those of higher rank. Enforced subordination should be regarded from this functional aspect.*

* Showing that an institution or structure performs a function does not justify its existence. War and terror have functions; we nevertheless try to find alternative ways of fulfilling them.

We also know from animal experiments that punishment inflicted by senior social partners, far from weakening the tie to them, tends rather to strengthen it. Low-ranking baboons take refuge with those of high rank even though the latter are the cause of their fear. Ducklings follow the mother duck or a dummy even more persistently if they are punished for it with electric shocks. Maltreated children are closely tied to the mother who maltreats them. Tyrants also use this binding mechanism (the terror bond). Finally, privations jointly undergone have a long-term binding effect on a group. The aim of initiation ceremonies—total identification with the group and the adoption of its rules and standards, which are partly preserved as a great secret—are assured by this means.* The phase of deprivation finally ends with ceremonial acceptance into the adult group.

The "Outsider" Reaction

Members of a group who depart from the norm in appearance or behavior often become a target of aggression. This forces the deviate individual to adapt himself again to the group norm; otherwise, the group will enforce his expulsion. Here aggression exercises a norm-preserving function, which is certainly advantageous for life in small groups in which harmonious coexistence is possible only if the behavior of other members is predictable. Everyone expects from others behavior in accordance with the norm, and deviation from it strikes him as surprising and strange—the first step toward estrangement and exclusion from the group. In general, the mechanism works blindly. The deviate individual may be mocked even if he has been disfigured or crippled by illness or accident, if he is too fat, or if he stammers and is not in a position to correct the "fault." This reaction occurs with brutal directness in schools; children can be extremely cruel in this respect. I remember an illustration that appeared in a UNESCO publication showing a small boy on crutches, with an embarrassed smile on his face, followed by a gang of boys throwing stones at him. As the same reaction has been observed among chimpanzees (see p. 66), we may conclude that it is a very ancient inheritance.

* Initiation rituals also exploit the fact that at the stage of puberty young people are at a sensitive phase in searching for values and are therefore ready to identify themselves with group values (Eibl-Eibesfeldt, *Der vorprogrammierte Mensch. Das Ererbte als bestimmender Faktor im menschlichen Verhalten* [Vienna, 1973]).

It follows a definite universal pattern, which I shall discuss later in the case of Bushmen. First of all, the behavior to which objection is taken is mocked and "aped." This shows the individual concerned what the offending behavior is and gives him an opportunity to adapt himself. If aping, mocking, laughter, and threats do not lead to a change, physical attacks occur.

Basically, these constitute a special form of educational aggression. Particular interest attaches to the ritualized forms of expression that precede the physical attack, such as showing the tongue, spitting, presentation of the buttocks or genitals, and laughing. The last of these seems to have the lowest release threshold. People laugh readily when other suffer minor mishaps, when their behavior diverges from the norm. They feel it to be "amusing"; in other words, it gives them pleasure. Certainly a child laughs at an absurd or clumsy movement only when it has itself mastered the behavior that it sees being carried out in a deviate manner.[20] The joke industry uses this readiness on our part by presenting us with laughter-releasing stimulus situations with the aid of simple dummies. There is a good market for aggressive humor, as the joke pages in the press show. Also, one laughs at someone who suffers minor harm as the result of clumsiness, hurts himself slightly, or breaks something. This *Schadenfreude* is shown even by children under the age of two.

Laughing at someone is certainly aggressively motivated, and it is indeed felt to be so by the person laughed at. At the same time, the collective aggressive reaction of laughter, which is infectious, has a binding effect on a group. Because collective aggression has a binding function, this form of aggressive behavior—laughter—becomes a signal of friendship in certain situations. Children laugh at play, and laughter is actually a sign of play. This contains the coded statement "Together we laugh at [threaten] a third party." There is a functional analogy to this in the animal kingdom, in which friendly greeting and courting ceremonies have repeatedly developed out of redirected threat behavior.[21] Laughter as a challenge to join in and laugh too—as a sign of friendship—also appears in situations in which a sudden release from tension takes place after the disappearance of a danger, after a slight fright, after an unexpected and surprising impression, and finally in certain situations resulting from the simultaneous working of antagonistic stimuli.[22] In all these situations, laughter seems to have the effect of releasing tension and diverting aggression. We shall return to this question later.

The connection between humor and aggression was seen by Freud. It can also be explained phylogenetically. According to van Hooff, laughing derives from the "relaxed open-mouth display" of monkeys, the latter being a ritualized display of an intention to bite, the "play face" indicating to the other party that the biting will not be real. Also there are homologues in the monkey world to the rhythmical sounds that are so typical of laughter. Rhythmical sounds generally uttered simultaneously by several members of a group serve to rouse feeling against a stranger to the group (mobbing reaction).[23]

A special form of norm-preserving aggression aims at adjusting differences in wealth between members of the group. This plays a big part in individualized small groups among primitive peoples. Those who possess something are envied and are exposed to aggressive pressures that compel them to share it. Envy, as I have mentioned, is a dominant motive in Bushman society. In this connection, Schoeck mentions the *muru* attacks that took place among the Maoris in New Zealand. *Muru* in the Maori language means "thunder." These attacks were directed against members of the community who were alleged to have committed a crime against it. Generally, however, this was merely an excuse. Anyone who owned anything worth looting could expect a *muru*. Any departure from the norm of everyday life, any expression of individuality, even an accident that made a man temporarily an invalid, was sufficient excuse for the community to attack the unfortunate and appropriate his personal possessions.

If a man's wife committed adultery, or his child died, or his grass fire spread over a burial place—any of these was sufficient excuse for a raid. The victim made no attempt to defend himself. This would have resulted only in his getting hurt, and would also have deprived him of the opportunity of taking part in the next *muru* himself. So he let himself be robbed by the community in the expectation of taking part in the next looting raid. Most movable property passed from hand to hand in this way and thus gradually became public property.

CONTROL OF INTRAGROUP AGGRESSION

Among the higher vertebrates it is extremely rare for aggression against a fellow member of a group to have fatal consequences. So far as I am aware, among anthropoid apes living in a state of nature, for example, there is no known instance of a member of a group being killed by a fellow member. The rare known instances of destructive aggression occurred in clashes with strangers to the group (p. 37). In the animal kingdom, the killing of conspecifics is prevented by innate inhibitions on aggression directed against the capacity to kill. The inhibitions are activated by a defeated animal's assuming a definite attitude of submission. Fighting is often ritualized. There are also other institutions that prevent aggression among members of a group from escalating and thus endangering the group's coherence. I have mentioned active intervention by those of higher rank who put an end to hostilities by threats or, if necessary, actual attacks, and finally, the loser's readiness to take his place in a ranking order. All these mechanisms are also active in man.

Man has at his disposal a number of ways of behavior that have an appeasing effect, inhibiting aggression and activating aid behavior. Among these are weeping, lamenting, lowering the head, pouting, and smiling in a friendly fashion, as well as a number of other expressive movements indicating readiness for social contact. The strategy of their use has not yet been sufficiently studied, but a comparative study of cultures shows that the same elements of behavior are to be met with in all of them, so far as is known, and that in particular situations, such as greeting, for instance, their use follows the same pattern.*

Children induce friendly feeling primarily by their appearance, and they are often used to signal friendship. Australian aborigines approached the feared white man by pushing children in front of them,[24] and in many cultures, including our own, guests are greeted by way of the child. At the palm-fruit festival of the Yanomami Indians, next to the warriors, who behave aggressively, children dance, waving green palm branches as a symbol of peace.[25] Television advertising also makes use of the aggression-appeasing effect of the child stereotype.

Similar use is made of women, who signal nonaggression, as is shown by a study by Baron. The ill temper of males angered for

* Schmidbauer argues that man has no innate inhibition against killing on the ground that it is known from various cultures that mothers kill unwanted babies immediately after birth "with a light heart" (*Die sogenannte Aggression* [Hamburg, 1972]). This would be convincing, in my opinion, only if the killing were really done with a light heart, i.e., without inhibition. Most anthropological reports indicate the contrary, however, and so does my personal experience with the !Ko Bushmen, the Yanomami Indians, and the Eipo of New Guinea.

experimental purposes was greatly reduced when they were shown pictures of sexually attractive nudes from *Playboy*; pictures with neutral content, such as landscapes, had no such effect.[26] Advertising often makes use of sexually attractive women. Until recently, I believed that these served only to catch the eye, but this study suggests that such pictures also have an appeasing effect.[7]

The offering of gifts, particularly food, serves the same purpose. These ceremonies are encountered in many cultures, and the spontaneous offering of food as a sign of friendship can be observed among young children. I suggested at one time that the friendly experiences associated with feeding might be the reason that people everywhere express friendship by offering food,[28] but later I preferred the idea that we are faced here with an innate behavior pattern, since quite small children resort to it spontaneously in appropriate situations, such as when they want to make friends with a stranger. There are also analogies in the animal kingdom: handing over food is built into courtship rituals as well as into others that have an appeasing function. It therefore surprises me that in a discussion of my derivation of peacemaking rituals, Horn, in an otherwise carefully considered analysis of the relevance of ethology to the human sciences, quoted only one of my examples. He refers to my account of a war experience of an acquaintance of mine who was inhibited in an attack on an enemy soldier when the latter spontaneously offered him a piece of bread. Horn describes this as a naïve argument, claiming that an adequate interpretation of the incident would require a thorough psychoanalytical examination of my acquaintance. "It can only be conjectured that to him the offering of bread (or of food in general?) had in the course of his life acquired a significance so contrasting to that of war that he could not deal with the situational contradiction except in this way, certainly in cooperation with the other partner."[29] In fact, I took this possibility into account, and it was more than this one example that led me to the assumption of an innate disposition.

Horn continues: "If Eibl-Eibesfeldt's interpretation is taken seriously, it would have to be possible to explain ethologically why no wars can be ended or avoided in this manner, or alternatively, on the basis of what human ethological laws other similar interactions end, as it were, in a normal warlike manner, and how it came about in the case of Irenaeus Eibl-Eibesfeld's acquaintance and his enemy that there was a happy breakthrough of what the ethologist calls primary nature."[30] I went into that too when I discussed the erection of barriers to communication and the dehumanization of the enemy in war, and I shall deal with this point in greater detail later.

In spite of the technical development of weapons, within the group the inhibitions on aggression activated by the signals enumerated above work very well. According to Russell and Russell, only 0.1 per 1,000 of the population of our Western cultural world die of murder.[31]

Ranking Order

In man, as in higher mammals, there is an obvious readiness to fit into a ranking order and temporarily to follow an individual of higher rank. Experiments by Hokanson and Shetler incidentally showed that annoyance (measured by increased blood pressure) lasted for a shorter time when the person causing it was of high rank. When the chief experimenter said he was only an assistant, the annoyance he caused in otherwise similar experimental conditions lasted longer.[32]

People are obviously willing to accept more from a high-ranking individual than from a low-ranking one. They are also more inclined to learn from a person of higher rank. In military training, a recruit is subjected to a breaking-in process aimed at fitting him into a low rank, thus making him willing to learn from and follow individuals of higher rank. An individual of low rank shows an evident tendency to obey authority, illustrated by Abraham's willingness to sacrifice his son, as well as by Milgram's experiments.[33]

Milgram conducted what purported to be learning experiments. Volunteers were told that the object was to investigate the effect of physical punishment on pupils' learning progress. Each volunteer was allotted a pupil, to whom he was told to administer an electric shock whenever he made a mistake; each successive mistake was to be followed by a bigger shock. The volunteers sat at an instrument that was purported to give shocks increasing in thirty stages from 15 to 450 volts. If the bigger shocks had really been given, they would have resulted in severe injury. Though the volunteers knew this, and were told at the outset that they could break off at any time, most of them followed the experimenter's instructions and continued to the very end, in spite of the protests and recorded cries of pain that were played back to them and their own protests against the experiment in which they were taking part. If the chief experimenter was deprived of his rank, the number of those who refused to continue increased, and if he gave his orders by telephone instead of in person, the number of those who refused to obey also increased. Many pretended to give bigger electric shocks without actually doing so, showing that no sadistic motivation was present.

These experiments leave open the question whether an innate or an acquired disposition was at work here. From the functional point of view, the effect was the same. Whether it is innate or the result of training, the disposition to obey prevents constant destructive struggles for rank, which would have a disruptive effect on the life of the group.* But as there is no reason to assume that there was a phase in the development from ape to man in which—because the latter lived a solitary life, perhaps—he was not subject to these functional compulsions, it is hard to see why the innate should have been eliminated and completely replaced by cultural adaptation in this instance. Also the previously mentioned findings of Stayton and his colleagues that children's obedience does not have to be enforced (p. 86)[34] points to an innate inclination to obey.

Ritualization of Conflict

Through the invention of various instruments that can be used as weapons, man became a potential killer; that is, under the influence of highly aroused emotions, he can kill a fellow human being, even one who is closely connected to him. As a cultural response to this capacity, man developed rules governing the use of weapons, of dueling in particular. The weapons are so used that the probability of killing one's opponent is diminished, and armed combat turns into a contest with fixed rules. Among the Walbiri and some other Australian tribes, men throw wooden spears at each other, aiming only at the legs. Yanomami Indians, acting in accordance with a fixed ritual, take turns striking each other's smooth-shaven lowered heads with long wooden poles. Generally, apart from

* The statement that they perform a function does not imply that ranking orders should be generally or uncritically accepted. We have already referred to the problems resulting from the aspiration for rank in anonymous societies. It is perfectly possible that better social models will be developed to meet the new conditions of a mass society.

lacerations, no serious damage is done. Similar examples can be multiplied to taste. Of course, both parties have to know the rules if these cultural controls on aggression are to work, and there must be a certain bond between them. But these conditions are often not present in conflicts between members of different ethnic groups, and even groups of common origin can erect barriers to communication and behave like strangers to each other. In these cases, weapons are used destructively. We shall deal with this more thoroughly in the chapter on war.

Verbalized Aggression

Certain patterns of invective and abuse must surely be the same in principle in all cultures. They imply degradation to the point of dehumanization, with the associated threat of exclusion. The other party is smeared with animal names (in Europe: pig, dog, monkey), or physical or social defects are attributed to him (cripple, coward, scum, cheat). Terms of abuse implying the breaking of an incest taboo are often used. In quarrels among the Mbowamb of New Guinea, for instance, physical defects are attributed to the other party ("You have breasts like a woman's!") or he may be ridiculed for his poverty ("You riffraff!") or neglect of his person ("You're covered with ashes, it makes me sick to look at you!").[35] All these terms are in essence very similar to those used in Europe.

Further, man can use words to describe a situation by saying such things as "Do you want to step outside and settle this?" thus challenging aggression, or by using a verbal stereotype as a releasing stimulus. Such statements are a challenge to respond in the same way, and this is done in formless abuse, in the playful teasing of a joking partnership, and finally in the highly ritualized song duel. Verbal stereotypes can take the place of action and of stimulus-releasing situations. This is a notable form of ritualization, and it certainly played an important part in the development of language. Perhaps the decisive selective pressure actually came from this direction.

The Eskimos of West Greenland practice the song duel. The two opponents stand facing each other. One of them sings, often striking a small hand drum, while the other tolerates this with apparent composure until his turn comes. Not content with ridiculing his opponent, the singer blows in his face and actually pushes him with his forehead. The other man replies with mocking laughter and resists the push with his own forehead. Thus the duel continues, each contestant singing in turn.

In his study of lawbreaking and its consequences among the Eskimos of eastern Greenland, König gives some examples of these dueling songs, among them, the following:

SONG DUEL BETWEEN THE EAST GREENLANDERS KOUNGAK AND ERDLAVIK

KOUNGAK: Let me too follow the women's boat, follow the boat and the singers as kayak man—though I'm timid by nature, though I'm humble by nature—as when I follow paddling a kayak—follow with the singing—no wonder he was glad—he who nearly killed his cousin —nearly harpooned his cousin—no wonder he felt pleased—was glad.

ERDLAVIK (*dancing*): But it only makes me laugh—it only amuses me—Koungak, that you're a murderer—and that you're so bad tempered by nature—so inclined to flare up—because you have no more than three wives—and you think that's too few—that's why you're so bad tempered—you should let other men marry them—then you'd get everything they caught—Koungak, because you don't bother about what others think—that's why you're always hungry—because your wives eat up everything—that's why you started killing other people.

IGSIAVIK'S DRUM SONG

(The other party is Misuarinanga, who is standing in for his stepfather, Ipatkajik, because the latter has run out of material for songs.)

"You think highly of him, you stick together with him. When you sing, you must put your arm around his neck, take care of him, and be kind to him." (He puts a peg in his opponent's mouth.) "I don't know what to do, since my opponent can neither sing nor let out a sound." (He puts a piece of wood into his opponent's mouth and makes as if to sew it up.) "What are we to do with my opponent? He can neither sing nor let his voice out. Since he has become inaudible, perhaps I should stretch his mouth and try to make it bigger." (He stretches his opponent's mouth with his fingers and uses a wooden stick to stuff some bacon down his throat.) "My opponent has a great deal to tell me: he says I wanted to do something to Akenatsiak and wanted to kill him. When we came from Stararmiut to Anitsuarsik from the south it was you who started a drunken dance with Akenatsiak." (He puts a strap around his opponent's mouth and ties him to a rafter with it.) "I don't know anything about having wanted to do anything to Akenatsiak or having wanted to kill him, nor do I know any reason why I should have done so. It must be because both of us were after his [Ipatkajik's] wife that you accused me of it. If he sings against me again, I'll sing again against him."

Whenever Igsiavik pauses between verses to perform various actions ridiculing Misuarinanga, the latter makes a show of indifference to demonstrate to the spectators that he is perfectly capable of talking and ridiculing him.[36]

By a remarkable parallel, similar songs have developed among the Tyroleans and other inhabitants of Austria. On social occasions in taverns—after a wedding or a cattle market, for instance—one man can recite mocking verses about another (the bridegroom, for example) with impunity. The performance takes the form of a playful public scolding, and it often discloses things at which other people have taken offense. But often two men can "duel" in this way. In the old days, this was frequently the introduction to a brawl. Von Hörmann describes how young Tyroleans would begin by singing against each other in the tavern, and how this would end up in a miniature battle; this, however, took place according to previously agreed rules and did not lead to bad feeling between the contestants. Here is an example of such a song duel:

Kimm vom Unterland aufer, Koa Weg ist mir z'weit, Hab' an Trager bei mir Mit'r Kraxen voll Schneid.	I've come up from the plain, No path is too long for me, I've a rucksack with me With a load of courage.
Bald's regnet, regnet's strichweis, *Bald's schneibt, so schneibt's kluag,* *Dein Singen geht stichweis,* *Das wert' i mir g'nuag.*	*Now it rains, rains in torrents,* *Now fine snow falls,* *Your singing goes in fits and starts,* *I'm beginning to have enough of it.*
Buxbaum und Ahornlaub, Türkisch Papier, Heut' hätt' i mei' glockspeisern Raufzeug bei mir.	Boxwood and maple leaf, Turkish paper, If today I only had My iron rake with me.
Hör' auf so z'singen *Und nachi schnagglen,* *I tät' mir dir schmeissen* *Und Fingerhagglen*	*Stop singing like that* *And cracking jokes,* *Or we'll be having a fight* *Or a finger-wrestling bout.*

Hör' auf a so z'singen	Stop singing like that,
Du spannlange Wurz',	You big beanpole,
Wenn d'abbrechen tätst,	If you broke,
Wärst zum Anknüpfen z'kurz.	You'd be too short to be put together again.
Dort oben auf der Heach	*Up there on the hill*
Steh'n drei Zunterbuachen,	*There are three beech trees,*
Wir wollen nit lang grein'	*Let's not squabble much longer*
Lieber d' Schneid versuachen.	*But try out our courage.*
A frischer Bua bin i	I'm a lively lad,
Han d'Federn aufg'steckt,	I've stuck a feather in my hat,
Im Raufen und Schlagen	In a roughhouse
Hat mi' koaner derschreckt.	No one has yet frightened me.
Büebel, wenn d'schlagst,	*Laddie, when you hit,*
Schlag grad nit auf d'Augen,	*Don't hit me in the eyes,*
Dass i no sehen kann	*And stop me from seeing well enough.*
Deine Scherben z'sammklauben.	*To pick up your pieces.*
Henneler, Henneler,	Henneler, Henneler,
Pack' nur gleich z'samm',	It's time to pack up
Dein hochmütig's Reimen	Your boastful rhyming,
Das dauert mir z'lang.	It's gone on too long for me.[37]

In this instance, Lüers, for instance, points out, the two young men were certainly thirsting for a fight. Normally, things did not go beyond the stage of singing rhyming songs. If the assailant became too crude, the assailed party, or a neutral, would sing a warning rhyme:

Tua net a so singa,	Don't sing like that,
Tua net a so sang,	Don't sing like that,
Sinscht tua-r-i da 'n Schnabl	Or I'll clean your mouth
Mit a Tennarauf'n zwang.	With an iron rake.[38]

According to Kochman, verbal duels also take place among blacks in the United States. One youth challenges another by making a derogatory remark about a member of his family. This compels him to reply, especially if others are present, and he makes a rude remark about the first youth's family, and thus the verbal duel gets into its stride; it generally ends without the participants coming to blows. These verbal duels play a big role, particularly in the fifteen-to-seventeen age group, and verbal agility is highly prized.

Boys and young men collect derogatory wisecracks, and those with the biggest repertoire win duels and gain prestige. Here are some examples:

> You mama is so bowlegged, she looks like the bite out of a donut.
>
> You mama sent her picture to the lonely hearts club, and they sent it back and said: "We ain't that lonely"!
>
> Your family is so poor the rats and roaches eat lunch out.
>
> Your house is so small the roaches walk single file.
>
> I walked in your house and your family was running around the table. I said: "Why you doin' that?" Your mama say: "First one drops, we eat."

The quickest-witted manage to turn verbal assaults back on their makers. Thus one youth responded to the phrase "Fuck you" by saying: "Man, you haven't even kissed me yet."[39]

Peacemaking, Consoling, Taking Sides (Mediating)

The behavior patterns of peacemaking and consoling, which play a big part in children's groups, are still relatively uninvestigated. Children spontaneously take sides with a child that has been attacked, and in a way that does not lead to an extension of the quarrel. They argue against continuing it and console the offended party by stroking or other cheering gestures such as offering him or her a toy or candy or getting the child to join in a game. I have often been struck by the fact that children as young as two react very strongly to simulated quarrels between their parents, and actually intervene and take sides. When a father, for instance, brusquely refused his wife's repeated requests to pass the sugar, their little girl of three promptly took the mother's part, alternating threatening motions with her hand with verbal appeals. Little is yet known about the ontogeny of this highly interesting pattern of interaction.

Let us bear in mind, then, that individualized human aggression is effectively held in check by a number of phylogenetic adaptations. In all cultures there is a marked inhibition against killing a fellow human being, and if it is desired to ignore it, as in war, for instance, special indoctrination is necessary if the sympathetic appeal of common humanity is to be disregarded. Sympathy as the subjective correlative of the inhibition on killing is felt in all cultures, and is everywhere released by the same signals. Thus inhibitions on aggression are innate in us. The commandment "Thou shalt not kill" is based on this constitutional factor. The invention of weapons facilitated murder and necessitated additional cultural patterns of control.

Binding Rituals

We human beings also have at our disposal other very effective counters to aggression in the repertoire of group-binding rituals derived from the parent-child relationship. A number of cultural-binding rituals, such as those of greeting and numerous feast rituals, are based on this very ancient heritage. I have shown elsewhere how important greetings are in avoiding friction in human social coexistence,[40] and we have already mentioned that greeting behavior and feasts everywhere exhibit the same construction and are also characterized by a number of universal behavior patterns.

These behavior patterns can be verbalized. An example is the stereotyped greeting formula of the Afghans, which begins with the "gift" of a good wish and goes on to express sympathy and concern. When two Afghans meet, this verbal ritual takes the same invariable course:

I wish you good health.	And I wish you good health.
How are you?	Thanks be to God, I am well.
You are well?	And how are you?
Are your children well?	Thank God, they are all well.
Are the other members of your family well?	Thanks be to God, they are all well. We hope your family are well too.
Thank you, they are all well. It is a long time since I saw you. Where have you been and what have you been doing?	I have had many cares. I have had no opportunity of coming to see you. I am still doing my old work and staying in Kabul.
With God's aid, will you be able to cope with your work and your worries?	Thank God, everything is going satisfactorily. And what are the prospects with you?
Not bad.[41]	

It would be a mistake to regard this as an exchange of empty banalities. Greetings perform an important function as lubricators of everyday social life. This also applies to feasts, which reinforce the bond between members of a group in a special way—at group level. While greetings smooth the path of individual relationship between a few persons, a feast has the effect of binding a larger group. There are family feasts, village feasts, tribal feasts, to mention only a few of the social levels at which they take place. Their structure is basically the same. A bond is established and reinforced by eating, exchanging gifts, conciliatory appeals, demonstrations of sympathy and concern, and joint activity such as dancing. The bond is further reinforced if aggression is diverted against third parties and unity is thus demonstrated against a common, often imaginary, enemy.

Shokeid studied the significance of social events in the life of an Israeli village. There were strong tensions between some of the inhabitants, who nevertheless met on friendly terms at ceremonies and entertainments. When villagers of lower rank issued invitations, there were many acceptances, as these individuals of lower rank were less concerned with village politics and acted as neutrals. This enabled bitter opponents to meet on neutral ground, and the relaxed atmosphere damped down their hostility. Shokeid was convinced that without such periods of detente, group life would be intolerable.[42] This certainly has a general application, and in many societies, feasts have been institutionalized. One may mention Christmas in this context. The emphasis on the exchange of gifts makes it clear that it is a "peace festival."

Safety-valve Customs

"Those who tease each other love each other." Aggressive intentions press for discharge. A number of customs that anthropologists very appropriately call "safety-valve customs" help to relieve the strain and enable aggression to be acted out harmlessly. So-called joking partnerships exist almost all over the world.[43] Joking partners are allowed to tease and often to be rude to each other, and thus to abreact their aggressions in a socially recognized way. Males and females of marrying age between whom joking partnerships exist are generally also potential marriage partners, while relationships between persons between whom sexual or marriage relations are taboo are generally highly formalized, and in these cases, joking relationships are not permitted. Among many primitive tribes, joking relationships exist between grandfathers and grandsons.

These relationships are especially relaxed and easy, as the grandfathers exercise no disciplinary authority; that is the father's business. In his relationship with his grandfather, the child finds relaxation from tension, for his relationships with other grown-ups are burdened by authority. An exception to this rule was found by Nadel among the Nuba, who live in extended families in which authority is exercised by the grandfather. In this case, there is a joking partnership between father and son.[44] This exception reinforces the view that joking partnerships constitute a safety valve.

Joking partnerships can also exist between groups. In Germany, a well-known example of this is the relationship between Bavaria and Prussia.

Aggression can also be acted out in sporting competitions. The wrestling bouts engaged in by Alpine peasants are well known. In the Bernese Oberland, it took the form of "swinging," or belt wrestling. Once a year, on a fixed day, young men from different valleys would meet in an Alpine meadow adjoining their home territories; men from Unterwalden and the Haslital, for instance, would meet on the Breitenfeld, and spectators would stream in from both places. The festival began with drinking at the inn; each contestant picked a man from the opposite side and drank with him. Then, to the sound of music, the two walked arm in arm to the arena. The contestants wore trousers with a special waistband that had a padded fold for gripping. They shook hands to demonstrate their friendly intentions. The object was to lift and throw one's opponent. The contestants stood or knelt opposite each other, each with his head on his opponent's right shoulder, holding his opponent's waistband by the fold. If a contestant was twice thrown on his back, he was beaten.[45]

At one time, contests of this kind took place all over the Alps. Wrestling bouts resembling the Swiss "swinging" bouts still take place in the Tyrol; they used to take place on borderline Alps such as the Hohe Salve in the Brixental. The contest would begin with young fellows holding each other's middle fingers and trying to force each other to relinquish their position; this is still done in the Tyrol and Bavaria. Then the wrestling proper began. The contestant who succeeded in throwing all his opponents was the winner and was allowed to wear their blackcock feathers in his hat.

The blackcock feather had great symbolic significance. When a young fellow felt like having a fight, he would stick one into his hat so that it pointed forward and go to the inn in the neighboring village, where its meaning was known. It did not take long until someone asked: "What does the feather cost?" The standard answer was "Five fingers and a grip," supplemented by an extemporaneous Alpine song. The other would respond with a similar quatrain, and so the appropriate combative mood would be stimulated.

Wearing a feather in one's hat also has provocative significance in Styria, the Salzburg area, and Bavaria, although the feathers of the ordinary domestic cock were used more often than those of the blackcock.

Comparable usage exists among many different peoples; nowadays, all over the world competitive games like football serve this safety-valve function.

Tensions within a society resulting from a rigid class structure, among other things, are also often acted out by means of safety-valve customs in which the oppressed are temporarily liberated by elevation to a top-dog role. On these occasions, they normally make fun of their masters, as is done in carnival orations in the Rhineland. People free themselves from tensions by mocking and scolding their masters and are then ready to accept the yoke again. Weidkuhn acutely describes carnivals as "polemics between social classes."[46] Cabarets perform a similar function, which is why, like political jokes, they are tolerated even by dictatorships. Aggression is acted out on a motor level in the form of laughter which, as we have already said, is a form of aggressive behavior. According to Oberem, Indian rural laborers in Ecuador dress up as landowners on festive occasions and imitate their behavior in mirth-provoking fashion, thus discharging their accumulated feelings of hatred.[47]

Finally, certain gift-exchange ceremonials have been transformed into rituals that serve to abreact aggressivity. The Goodenough Islanders overwhelm their enemies with gifts of tubers and pigs to shame them and triumph over them morally.[48] The same applies to the Kwakiutl potlatch custom (see pp. 85 and 130).

INBORN PROGRAMS IN HUMAN AGGRESSIVE BEHAVIOR

Releasing Stimuli

Even among infants, physical pain releases the reactions of flight, defense, or counterattack. This must be universal and phylogenetically very ancient, for pain is a powerful aggression-releasing stimulus among animals also.

Infants who have not yet learned to talk will attack people who take something away from them or refuse to do what they want. I have observed in very different cultures that infants strike their mothers when they fail to give them the breast quickly enough or do not give them a share of some tidbit. They also protest if they are interrupted at play. Every interruption of an enjoyable activity, every obstacle in the way of a goal-directed activity, every frustration of a wish, first of all rouses aggression. That applies universally to us human beings and is the basis of the frustration theory of aggression (see p. 78). It can be regarded as proved that experiences of deprivation (frustrations) result in acts of aggression which aid man to achieve his aims.

The reaction is not to a simple key stimulus, but to a more complex stimulus situation. I have tried to describe it by the term "situational cliché." We must assume that there are detectors (innate releasing mechanisms) that are tuned in to definite situations characterized by definite patterns of personal relationships and processes. Aid behavior, for example, is released both by signals, such as the crying of a child in distress, and by the situation itself. Similar considerations apply to the situations we have described that release jealousy for the defense and acquisition of places and possessions, whether physical or mental (ideas).

Our reaction to strangers is worthy of note. In all cultures, they are met with a certain reserve. Fear and rejection of strangers develop even in the absence of bad experiences with them. A three-month-old infant smiles in friendly fashion at every stranger, but at nine months, the sight of a stranger will make him freeze, and after a short time, he will begin to cry. He is obviously afraid. This phenomenon has been interpreted psychologically as fear of separation. Spitz says that the child compares the stranger's face with its memory picture of the mother (the libidinous object), and concludes: "That is not my mother, she is never coming back, I have lost my mother."[49] But this interpretation conflicts with Spitz's own observation that a child will show fear of strangers even in its mother's arms.

In the course of the child's development, this fear of strangers develops into a less fearful rejection of them, which can finally be marked by aggressive traits. A child will strike at a stranger who seeks contact too pressingly. Strangers activate the intraspecific agonistic system (flight and aggression).

I have observed rejection of strangers even among children born blind and deaf who had certainly had no bad experiences with strangers, since everyone was at pains to treat them kindly and give them a sense of security. These children had at most experienced punishment at the hands of those responsible for their care. Yet they distinguished strangers from acquaintances by smell, turned away from the former, or even showed their rejection by striking out at them. This was especially evident in the case of a girl who had been lovingly brought up by her mother and was later treated just as lovingly by her teachers in an institution. I observed rejection of strangers by five other children born blind and deaf, and I have seen the same phenomenon in all the cultures I have visited.*

The fear-of-strangers reaction applies quite specifically to strange conspecifics. Young children also show fear of strange animals, but they seldom show the often panicky fear that afflicts them in the presence of human strangers. If they once get to know a dog, they will try to stroke every strange dog they meet, even if it looks quite different. Children certainly do not apprehend human strangers as totally different and alien creatures, but as fellow human beings who are perceived as strange and feared as such.

It is this fear and rejection of strangers that underlies our tendency to form closed groups and react aggressively toward strangers who try to enter them. In such cases, strangers can activate collective aggression (group defense).

Strangeness is, of course, not in itself the key stimulus in this pattern of reaction. Instead, we must assume that conspecifics are the bearers of aggression-releasing signals. We know, for instance, that staring eyes are felt to be threatening.† These signals do not seem to have any effect on the infant in the first few months of life, but later, when the capacity to react to them has matured, their effec-

* The fear-of-strangers response in the !Kung Bushmen has been studied by M. J. Konner ("Aspects of the Developmental Ethology of a Foraging People," in *Ethological Studies of Child Behavior*, ed. N. Blurton-Jones [Cambridge, 1972], pp. 285–304). For a recent discussion of the phenomenon, see L. A. Sroufe, "Wariness of Stranger and the Study of Infant Development," *Child Development* 48 (1977): 731–46.

† More about eye contact and its consequences in M. Argyle and M. Cook, *Gaze and Mutual Gaze* (Cambridge, 1976); and R. G. Coss, "Eye-Like Schemata: Their Effect on Behaviour" (Disseration, University of Reading, 1972).

tiveness in relation to those with whom the child is most intimately associated is suppressed by the bond of personal knowledge. Thus fear of strangers is really the fear of human beings who have not been covered by friendly experiences. The fact that strange behavior releases mild forms of aggression (laughter) fits into this context. Those who behave divergently alienate themselves to a certain extent from their associates.

The detailed characteristics responsible for the fear of others are unknown to us. There are sufficient indications that innate threat signals are present in facial expression and bodily attitude. But since a stranger releases fear in a child even though he is making a friendly approach to it, there must be other characteristics inherent in us that activate the agonistic system and release flight or attack if the critical distance is overstepped.

Phylogenetic Adaptations in Motor Activity

There are behavior patterns in man corresponding to many of the threatening and fighting movements I have described in the chimpanzee. We too stamp our feet when we are angry, strike surfaces with the flat of our hands, and take a threatening posture in which we turn our arms inward. On these occasions, we actually raise the fur that we no longer possess; we feel the contraction of the hair-raising muscles as a shudder that runs over us. If we had fur, this would result in a striking enlargement of our apparent bodily size. Those resorting to intimidation behavior generally also use cultural means (clothing, adornments) to increase their apparent size, for instance, by emphasizing their shoulders and wearing head ornaments, while the submissive do the opposite: they make themselves look smaller. Besides inborn motor patterns, innate releasing mechanisms play an important role in determining human display behavior.[59] Both primates and man also threaten with objects (sticks, etc.) in basically the same way (see pp. 75 ff). Over and above these, there are a great many specifically humans forms of expressing agonistic behavior that are universal and therefore probably innate. Among these are the threatening stare and the other party's giving in by looking away, dropping the head, and pouting. We have mentioned that as early as the crawling stage, children push each other over, scratch, strike with the hand, and bite. As evidence of the existence of innate movements in the service of agonistic behavior, we can also cite the behavior of children born blind and deaf, who show all the important elements of threat behavior when they are angry. For instance, they clench their fists, stamp their feet, frown, bare their clenched teeth, and in extreme agitation will bite their own hand.

For the sake of completeness, I should like to mention a very old pattern of threat and intimidation behavior, though I shall do so only briefly here as I have dealt with it in detail in my book *Der vorprogrammierte Mensch*. This is phallic threatening, derived from a sexual domination gesture (mounting), which has been described among many monkey species in both the Old World and the New.[51] Among guenons and baboons, a few males always sit with their back to the group keeping guard, displaying their strikingly colored penis and their sometimes similarly strikingly colored testicles. If a stranger to the group approaches too closely, the guards actually have an erection; so-called "rage copulations" also take place. It is a striking fact that in the most varied cultures, phallic threatening appears in carved figures (amulets) intended to ward off evil spirits and protect persons and property. There are also numerous observations of mounting being carried out as a direct demonstration of domination.

Motivating Mechanisms

Is there an aggressive drive and consequently an appetence for aggressive clashes that can be abreacted? The answer is certainly yes. Many people are aggressively motivated at times. They seek out clashes, and opportunities for acting out aggression in ritualized form are often taken advantage of. One only has to think of combative games. Safety-valve customs of that kind are to be found all over the world, even in cultures whose ideals are thoroughly pacific.

Also it can be shown experimentally that aggressive tensions (accumulated aggression) can be acted out, "abreacted," we say, by aggressive acts. Hokanson and Shetler noted a substantial increase in the blood pressure of students whom they deliberately angered. The students were then divided into two groups. The experimenter told them he was going to solve problems, and asked that whenever he made a mistake, they let him know by pressing a button. One group was told that this would give him a punitive electric shock, and the other that he would merely see a blue light.

The blood pressure of those who believed they were administering electric shocks, that is, that they were acting aggressively against the experimenter, fell very quickly, and subjectively, too, their anger had subsided after the experiment. But the others, who by a similar action thought they were only switching on an electric light, remained angry for a long time.[52] Other experiments have shown that engaging in verbal aggression and watching films with an aggressive content have a tension-discharging effect.[53]

More recent experiments showed that aggressive and nonaggressive humor reduced the anger of the individuals concerned, and there is evidence that aggressive humor is rather more effective in this respect.[54] Landy and Mattee reported that looking at comic drawings of aggressive and nonaggressive content reduced verbal aggression against an experimenter who had previously angered them.[55] This suggests that humor has a cathartic effect.

The work of Baron and Ball shows that nonhostile humor reduces aggression. These authors regard it as inappropriate to assume a specific discharge of aggression in this case; they prefer to attribute the change of mood to the arousal of emotions that work against aggression. They suggest that, by being made to laugh, the subjects of the experiment were put in a frame of mind that was inconsistent with the aggressive actions they were asked to carry out immediately afterward—administering punitive electric shocks in a simulated learning experiment.[56] What took place here might well be a change of mood resulting from the activation of mechanisms that run counter to the aggressive system. In the chapter on the control of aggression, I mentioned that aggression can be appeased by friendly appeals. But Baron and Ball's experiments did not show conclusively that that was the case. Since laughter, as I have observed, is primarily aggressive, motor completion could lead to the abreaction of accumulated aggression even if the laughter was released by surprise and not by the aggressive content of the joke.

Some authors doubt whether games, television violence, and aggressive humor have a tension-releasing effect, asserting rather that they stimulate aggression.[57] I have repeatedly pointed out[58] that evidence that an aggressive film, for instance, has a tension-releasing effect should not mislead us into promoting such films as a cure for aggression, because we must assume that an individual who is not in an aggressive mood would be made aggressive by them. Even if a film enables aggression to be acted out, it must be borne in mind that the repeated activation of a physiological system leads to the training of that system. That is why it is naïve to base educational programs on the theory that children must be allowed, or even encouraged, to follow up and live out their aggressive impulses if they are to grow up into peaceable adults. As in the case of every drive, the cathartic effect, the release of tension, is temporary. On the other hand, it would be wrong to deprive children of aggressive experiences, because without such experiences, they would have great difficulty in dealing with and socializing their aggressive impulses.

There has recently also been lively discussion of the role of games. While some regard them as a valuable safety valve for the acting out of aggression, others regard them as dangerous, because they provide training in aggression. Long-term and short-term effects are again confused here. Many of the authors concerned are ignorant of even the most elementary physiological connections. Sipes, in his attempt to refute the catharsis theory of aggression, starts by assuming that, if the theory were correct, combative games should be less frequent among warlike than among peaceful peoples, for the warlike ones would abreact their aggression differently. He also assumes that in time of war, according to this theory, the practice of aggressive sports would decline.* He found that none of these "expectations" was confirmed. Warlike cultures actually practice more combative sports than unwarlike cultures, and in the United States, there was no diminution in combative sports during the Second World War, the Korean War, or the war in Vietnam.[59]

The first comment to be made is that no ethologist would have expected anything else. Warlike societies train for aggressivity, and consequently they practice combative sports. These provide the group with long-term training in aggression as well as a short-term

* "The Drive Discharge Model predicts somewhat similar levels of aggressive behavior in all societies, although the mode of expression can vary. It predicts an inverse relationship between the presence of war and of warlike sports in societies, which we should find expressed in two ways:

1. "An inverse synchronic relationship should exist between societies, with more warlike societies less likely to have (or need) such sports and less warlike societies more likely to have these sports.

2. A diachronic relationship also should exist within a given society, with periods of more intense war activity accompanied by less intense activity in warlike sports and periods of less intense war activity associated with more intense sports activity" ("War, Sports and Aggression: An Empirical Test of Two Rival Theories," *American Anthropologist* 75 [1973]: 64).

catharsis. Safety-valve mechanisms in the form of games, song duels, etc., also occur in peaceful cultures, but Sipes overlooks this.

The Bushmen, for instance, whom Sipes puts in the category of peaceful cultures lacking in combative sports, have many kinds of game, such as spear throwing, and a number of dances that are highly competitive.[60] They also hunt. Sipes calculates the increase in hunting activity in the United States (measured by the number of licenses divided by the number of potential hunters) in order to show that war does not lead to a decline in combative sports.* He classifies hunting as a combative sport, obviously without noticing that in his list of peaceful societies that he imagines to have no combative sports he includes many hunting tribes. Apart from this and other quite basic illogicalities that make his argument almost completely worthless, I would not, as I have already said, expect warlike cultures to be less inclined to warlike sports. In view of these errors in method and theory, his ambitious conclusion that we should throw the human aggressive drive model overboard is disturbing:

> War and combative type sports therefore do not, as often claimed, act as alternative channels for the discharge of accumulable aggressive tensions. Rather than being functional alternatives, war and combative sports activities in a society appear to be components of a broader culture pattern.
>
> However, the Drive Discharge Model is so entrenched in Western science that there should be investigation of other activities which conceivably could act as alternatives to war (and now, as we have seen, to combative sports as well) in the discharge of postulated drive tensions. Likely candidates are suicide, murder, punishment of deviants, drug use, physical assault on family or other community members, gossip, psychogenic illnesses and malignant magic. Unless there are definite indications that they serve as alternatives to war and combative sports, we can set aside the Drive Discharge Model with full confidence that it is not applicable to humans.[61]

It is as simple as that!

In his criticism of the catharsis hypothesis, Michaelis says that there are really two hypotheses which should be differentiated. The

* Incidentally, I would not directly equate hunting with aggressive behavior against conspecifics in this manner. I have mentioned that the two often depend on different physiological systems. This may not be the case in man, however, since no clear dividing line seems to exist in his case or in that of his closest relative, the chimpanzee (see p. 74).

first dates back to Aristotle, who believed that man was purged of his emotions by the exaggerated representation of his problems in tragedy. This theory was taken up by Freud, who, in accordance with his theory of drives, expected internal tension to diminish after an act of aggression was carried out. But according to the second catharsis hypothesis, which is supported by Lorenz, after aggressive activity, the probability of carrying out further aggressive acts declined.[62] My view is that these different ways of putting it actually describe the same phenomenon from different aspects. Freud was writing about subjective experience and Lorenz about observed behavior.

According to this, diminution of inner tension should reduce the probability of further aggressive action, but an experiment by Berkowitz seems to contradict this. Individuals who had been angered for experimental purposes, and had been able to give electric shocks to the person who had angered them, felt freer of tension, but they still had a worse opinion of the experimenter than did others who had been provoked by him but had not been able to get their own back.[63] The explanation of the inconsistency may be that the angered individuals, while abreacting their anger by their aggressive behavior, simultaneously reinforced their hostile image of the experimenter and expressed this in their subsequent judgment of him. A negative judgment cannot automatically be equated with aggressive behavior.

Certainly there are a number of variables that influence the catharsis effect. Hokanson, for instance, found notable differences between the sexes. He put two persons of the same sex, one of whom was an accomplice of the experimenter, in separate rooms. They were in communication by telephone, and each could react to the other's behavior by reward (indicated by a blue light) or punishment (administering an electric shock) or they could react neutrally. The experimenter's reactions were always programmed in advance. It turned out that the blood pressure of both males and females rose sharply after they were given an electric shock by the other party (the experimenter's accomplice). In the case of males, it dropped rapidly after counteraggression, but not when they rewarded or reacted neutrally. Females, however, reacted in the opposite way, with a quick drop in blood pressure when they rewarded but not when they reacted aggressively or neutrally.[64]

A similar experiment among prisoners showed that those who,

according to the records, generally reacted to threats with aggression behaved like the males in the experiment we have just described, while those who generally reacted passively followed the female pattern. This may be attributable to hormonal differences, or to differences in the learning history of the individuals concerned.

Learning experiences certainly play an important role. In another experiment, Hokanson and Shetler used a different set of arrangements: the experimenters gave females an electric shock whenever they reacted in a kind fashion; whenever the females applied shocks they received a friendly response. Exactly the opposite schedule was applied when the experimental subjects were males. The males grew friendlier in the course of the experiment and the females more and more aggressive, until in the end, the females were showing a reduction of tension after aggression and only a slow decline in tension after kindness, while the male reaction was the exact opposite. Subsequently, when the experimenters again reacted "haphazardly"—as in the original experiment—the original pattern of reaction reestablished itself. The experimenters concluded that obviously all reactions that, according to the experiences of the individual concerned, lead to the ending or avoidance of future aggression can cause a diminution of tension.[65] This does not conflict in any way with the concept of drives. Tension can be resolved by appeasement (friendly behavior) or by abreaction. That is completely in accordance with ethological theory. In animal groupings, those of lower rank often accept their position without any demonstrable signs of stress and react in accordance with their role. The different primary-reaction patterns of male and female may depend on inherited, hormonally determined, or sex-specific disposition, but are modifiable through individual experience.

Buss and Feshbach found that aggressive acts lead to a catharsis only when accompanied by "anger affect." "Cold" aggression, carried out without emotional involvement, actually increases the probability that further acts of aggression will follow.[66] Tensions produced by provocation are resolved if the necessity of the provocation is adequately explained[67] and is consequently felt to have been justified. Similarly, aggressivity fades if a third party ensures that the provocation ceases; the person provoked is best appeased if the intermediary tells him that the provoker has been punished.[68]

Recently a number of studies have confirmed that the opportunity to "hurt" an insulting person physically tends to reduce the angered person's subsequent aggression against him.[69] In contrast to these findings, Geen, Stonner, and Shope reported an increase in aggression under apparently similar conditions.[70] In reply, and on the basis of further experiments, Konečni and Ebbesen report that the failure of Geen and his coworkers to repeat the findings of his group may be partially due to a weak induction of anger in the experimental subjects. Insulted subjects show cathartic effect when allowed to act aggressively against an annoyer.[71]

Controversy nowadays centers not so much on the question whether an aggressive drive exists—few have doubts about the dynamics of aggressive behavior—as on whether the drive systems that cause combativeness are acquired in the course of the ontogenic development (the secondary-drive hypothesis) or are innate in man (the primary-drive hypothesis). Those who maintain the former assume that aggression helps other drives and is always activated when these are repressed. Arno Plack, for instance, traces all aggression back to repression of the sex drive, and claims that there would be no aggression in the event of complete fulfillment of the primary drive.[72] But that is speculation. The fact that frustration of every kind encourages aggression does not prove that frustration is its only cause.

There is no conclusive proof of the existence of a primary aggressive drive, but there is strong circumstantial evidence that suggests it. I am referring not so much to the universal dissemination of aggressive behavior and the presence of safety-valve customs in peaceful cultures, for there are no cultures in which deprivations do not occur in some form or another. Aggressive behavior in childhood everywhere leads to success, and the assumption that aggressive behavior is trained in this way would certainly seem to be indicated by the principle of economy. In addition to these manifestations of secondary drives, neurological findings point to the existence of primary drives in the central nervous system. Gibbs, Moyer, and Sweet and colleagues have demonstrated spontaneous outbursts of rage of neurogenic origin in man. They are accompanied by typical electrical brain activity, and rage can actually be reproduced by electrical stimulation of the regions concerned. The rage so caused expressed itself in spontaneous attacks on individuals, the smashing of objects, the facial expression of anger, and subjectively in feelings of anger.[73] The anger centers lie in the temporal lobes and amygdaloid nuclei, whose spontaneous activity results in patients' rage. As it is known that healthy individuals also possess these structures, and as it is also known that there are no ganglion cells that do not show spontaneity,[74] it can be assumed that combativeness normally depends on, among other things, the normal spontaneous activity of these centers. It is also known that they are normally kept under control by other centers. Violent patients can be calmed down by stimulation of the ventromedial portion of the frontal lobes and the central region of the temporal lobes. Patients have had electrodes and self-stimulation equipment permanently implanted to enable them to suppress rage when they felt it rising. This method is probably better than the common practice of complete operative removal of the rage-releasing areas of the brain, which certainly helps but leads to a large number of undesired concomitant phenomena, since many other centers are affected by the operative intervention. In animals, incidentally, when the inhibiting centers have been destroyed—by removal of the cerebral cortex, for example—spontaneous fits of rage occur, showing that the spontaneity of the aggressive system manifests itself if the normally inhibiting influence is removed. In this connection, I would recall what I said earlier about paradoxical sleep (p. 58).

LEARNING PROCESSES IN THE DEVELOPMENT OF AGGRESSIVE BEHAVIOR

Learning as well as maturation processes play a large part in the development of aggressive behavior in animals. We have seen that mice can be trained to be aggressive by successful fights, even if this leads only to the driving away of conspecifics, and that they can also be trained to be peaceful by being defeated or punished in some other way for aggression. If this is the case among relatively low mammals, we are entitled to expect experience to play an important role in man. In the development of aggressive behavior, a major role is played by learning from success, learning from social models, and finally by deliberate educational policy making use of rewards and punishments.

Bandura and Walters tested the effect of aggressive models on children. One group was shown an adult maltreating a rubber doll. The second group saw the same thing on a television screen, while a third was shown a cartoon film in which the same aggression was carried out by a cat. A fourth group was shown no aggressive model. Then, when all four groups had been frustrated and their play was observed, it turned out that the three groups that had seen an aggressive model were notably more aggressive than the control group. Those who saw the aggression on the screen behaved more aggressively than the first group who saw it in reality.[75]

A further experiment by Hicks showed that there were long-lasting aftereffects. He experimented with a group of children just as Bandura and Walters had done, and when he tested them again six months later, the effect was plainly discernible, though only among the children who had seen the adult model.

Boys who were shown a film with a saber-fencing scene behaved more aggressively than others who saw a film of the same length but with neutral content.[76]*

These examples should be sufficient to show how important learning from models is in the development of aggressive attitudes. Like many other more recent studies,[77] they also provide evidence of how aggression is encouraged by violence seen on television. Those who have demonstrated that such scenes have a tension-releasing (cathartic) effect and consequently believe they have proved the innocuousness of representations of violence are focusing on the short-term effect. They overlook the long-term impact, which consists of an increase in aggressivity similar to that which takes place in various forms of combative games.

Learning by social imitation certainly plays a very important part in the normal development of the child. Without any pressure from its educators, and entirely on its own initiative, the child identifies with the parent of the same sex.

On top of this, education by adults has a directly formative influence on the child's attitudes. Among warlike peoples—a category that includes us Europeans—boys are taught not to take things lying down and to respond to aggression with counteraggression. The same applies among the Himba, a nonacculturated Herero tribe in the Kaokoveld in South-West Africa (Namibia). I filmed an incident there in which a boy was beaten by another boy and ran howling to his hut. His grandmother gave him a stick and told him to go and beat his assailant with it. He did not dare do this and wept more loudly, whereupon his grandmother boxed his ears and left him lying in front of the hut. I collected many examples of the

* The boys' aggressive statements and the number of electric shocks they administered in a simulated learning experiment were recorded.

same sort of thing among the warlike Yanomami Indians for whom self-defense is a moral duty. Adults see to it that this principle is adhered to even by young children, girls as well as boys. A weeping girl who had been struck by her brother was given a stick by her mother to hit him with. Since he was the bigger and stronger of the two, his mother held him. She also showed the little girl how to bite the boy, and encouraged her to do it.

I have many films that show mothers not just encouraging their young children to get their own back, but inciting them to aggressivity in general by teasing and angering them. When the children thus roused finally attack their tormentors, the adults just laugh. Chagnon reports how on the occasion of a Yanomami feast

> all the young boys from eight to fifteen years old [were made] to duck-waddle around the village periphery and fight each other. The boys were reluctant and tried to run away, afraid they would be hurt. Their parents dragged them back by force and insisted that they hit each other. The first few blows brought tears to their eyes, but as the fight progressed, fears turned to anger and rage, and they ended up enthusiastically pounding each other as hard as they could, bawling, screaming, and rolling in the dirt while their fathers cheered them on and admired their ferocity.[78]

In addition to the fact that combativeness and savagery among the Yanomami are actively encouraged by education, Yanomami children identify themselves in play with the adult roles expected by their culture. They shoot blunt arrows at each other, imitate the adults' intimidation dances, and take turns striking each other on the head with clubs made of soft wood, following the rules governing adult duels (p. 95).

Socialization in a culture with peaceful ideals is very different. Among the Bushmen of the Kalahari, I have never seen a child encouraged to counterattack an aggressor. Only very small children are allowed to hit each other playfully with sticks, which causes amusement. Crawlers who go for one another are separated and pacified. As soon as children can walk, they join play groups in which further socialization takes place. Aggression is not tolerated by the older children, who intervene, scold, and even strike an aggressor, and pacify and console the victim. They give instruction in sharing things and in joint play, and thus reinforce behavior patterns that work against aggression. There are competitive games in which two sides, for instance, compete in throwing sticks at pieces of wood stuck into the ground and trying to knock them down, but these games are highly ritualized and are not direct preliminary training for real combat.[79]

Finally, learning from success plays an important part in the formation of aggressive behavior patterns. Children quickly learn that they can get their way by aggressive protest, and it is to be assumed that in many cultures, they collect these experiences and use aggression instrumentally to overcome obstacles in the way of what they want. I have mentioned that children use aggression exploratively. When their demands meet with no resistance, learning from this success leads to an escalation of demands.[80]

No ethologist has ever doubted that human aggression is to a large extent determined by individual experiences. Michaelis has suggested a functional model of aggression that seems especially well adapted to illustrating the possibilities of modification of aggressive behavior in man (figure 11).[81] It plainly shows the interaction of constitutional and learned components, of inhibiting and encouraging influences. The model shows a behavior sequence, a chain of binary (yes or no) decisions in a control circuit. When the organism becomes aware of the approach of a conspecific, it has to decide whether or not to regard this as a disturbance.* If the individual is outside its territory, this is not felt to be disturbance, but it is felt to be one if the boundary of its territory is overstepped, and the organism is activated in accordance with the intensity of the disturbance. In the first place, it will check whether ways and means of eliminating the disturbance (the discrepancy between what is and what should be) are available to it.

If no such possibility is available (for instance, if the organism is unable to move), no action follows. But if there is a choice among several kinds of behavior, the one chosen, according to Michaelis, will be that offering the greatest likelihood of success, in other

* In this connection, Michaelis speaks of disturbances of a homeostasis and defines aggression as behavior intended to preserve or establish a homeostasis at the expense and against the resistance of another organism; in this context "homeostasis" means "subjective well-being." The subjective element in this definition limits its applicability. In man, of course, the presence of discomfort felt at the approach of a stranger, for instance, could be established by questioning. But we can accept the definition if we equate disturbance of a homeostasis with disturbance of a state of affairs that the organism, by reason of its experience and phylogenesis, seeks to maintain or attain.

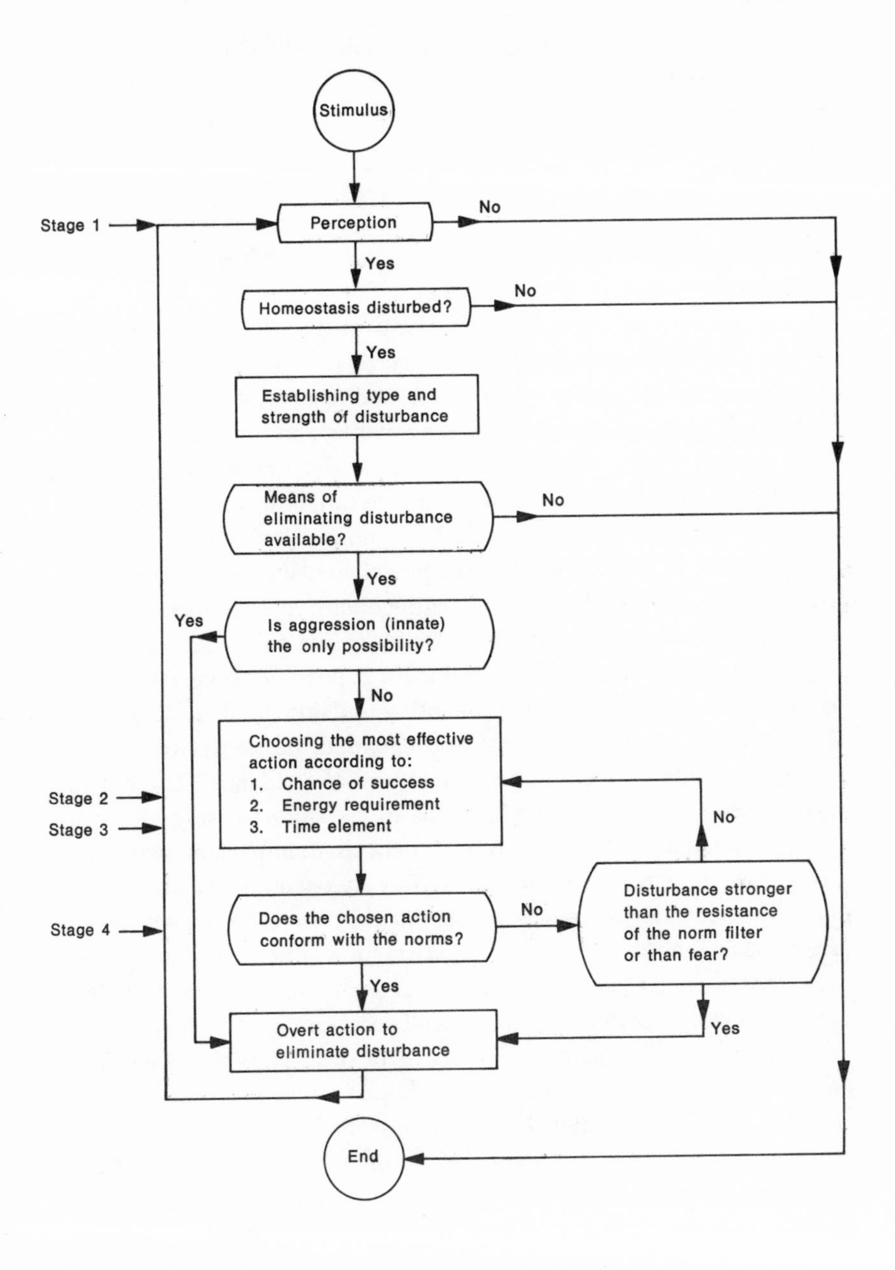

Figure 11. Theoretical model of the functioning of aggression, with possible starting points (Stages) of externally guided behavior changes. From Michaelis (1974).

words, aggression, which is the most effective. Our observations certainly teach us that threatening is the first form of aggressive behavior used, and that physical aggression takes place only if threatening is unsuccessful. This pattern of behavior is subjected to a control by norms. If it passes the norm filter, it takes place, according to Michaelis, with an intensity that corresponds to the extent to which the homeostasis has been disturbed.* After every action, testing takes place to find out whether the disturbance still exists (arrow at the beginning of each chain of events). If the pattern offering the highest probability of success (for instance, killing the intruder) cannot pass the norm filter, the next most effective is examined. If only a single pattern is available, the norm filter, which works as an inhibitory mechanism, alone decides whether or not aggressive action ensues. The disturbance can, however, be so strong that it overcomes the resistance of the norm filter, with the result that aggression as the ultimate arbiter takes place even if the norm control says no. In this system, endogenous motivations and the influence of antagonistic drive systems are not taken into account.

Examination of the model also makes it possible to visualize possibilities of guiding behavior by other outside influences. In the first place, if no disturbance is apprehended, no aggressive action takes place. Consequently, one possibility of guidance (Stage 1 in the figure) would be to diminish the aggression-releasing effect of the incoming stimuli or to dismiss them as unimportant, with the result that no action would follow. But that would not always be possible. Michaelis correctly points out that there are needs (food, sleep, sex) that always press for satisfaction. But it is equally certain that there are secondary needs of lesser importance and that guidance of needs in this field, as well as of the means of satisfying them, could act as counters at this level of aggression. Control is also possible by the manipulation of key stimuli (inherited and acquired). Needs affected by learning influences can cause disturbances when there are obstacles to their satisfaction. But if the individual sees that their satisfaction is out of his reach, in other words, that no ways and means of successfully achieving satisfac-

* Michaelis assumes that aggression is always purely reactive, which is certainly not in accordance with our views. We would also introduce a spontaneous component, but it is unnecessary to burden the model with this. The steps in the process of decision and the possibilities of action can be visualized without it.

tion are available to him, he will put up with the situation for the time being. In keeping with this, revolutions apparently do not take place when hardship and oppression are at their worst, but only when a phase of deprivation is followed by one of prosperity and is then followed by another phase of hardship (such as that resulting from a recession).

One way in which learning processes can help to keep aggression under control is by the individual's learning to use nonaggressive patterns of behavior. If a society fulfilled wishes in response to requests instead of in response to aggression, aggression would hardly continue to be used to secure their fulfillment (Stage 2 in the figure). Man can also learn that needs do not have to be satisfied immediately. Prolonging the time span pending fulfillment of the wish may be more difficult, though education normally works effectively in this direction.

The fourth possibility of modification, placed at the end of the chain, results from influencing the inhibitory mechanisms. The inhibitions against aggression can undoubtedly be reinforced by education. In fact, it is here that socialization normally begins. The child finds out which forms of aggression are justified, and therefore permitted, in various situations and which are not. Innate inhibitions can be encouraged or actually eliminated in this way.

6
Intergroup Aggression and War

Animals and birds fight for the possession of females or the defense of hunting grounds or for leadership of the herd, but it is said that they do not fight each other to the death as men do in war. . . . These facts indicate that the human institution of war is more than the human form of the aggressive drive that is inherent in every animal. War must be a human aberration or aggravation of the aggressive drive. It must also be a product of tradition and not of instinct.

ARNOLD TOYNBEE[1]

THE CULTURAL EVOLUTION TO WAR

In discussing the functional laws of phylogenetic cultural evolution, I mentioned that, under the formative influence of similar selective pressures, cultural evolution copies biological evolution at a higher level of the development spiral. The formation of species, therefore, has its counterpart in cultural pseudospeciation. Cultures mark themselves off from each other as if they were different species, and so adapt themselves to different niches. To emphasize their difference from others, representatives of different groups describe themselves as human, while all others are dismissed as nonhuman or not fully equipped with all the human values. This cultural development is based on biological preadaptations, above all, on our innate rejection of strangers, which leads to the demarcation of the group.

Group identity is brought about by aggression; with it, the group assures itself of territory, which it collectively defends against strangers. Up to this point, there is no basic difference between human conditions and those we encountered among the chimpanzees. But the possession of weapons enables men to kill one another.

Weapons were certainly a vital factor in the development of destructive aggression. Weapon techniques have to a certain extent circumvented our innate inhibitions. A rapid blow with a weapon can eliminate a fellow human before he has a chance of appealing to our sympathy by an appropriate gesture of submission. This is even more likely to be the case if the killing is done from a distance, for instance, with an arrow. Nevertheless, we must bear in mind that, in spite of the possession of weapons, intragroup aggression seldom becomes destructive. But intergroup aggression—war—is essentially destructive, and must therefore have some other basis. The vital role is played here by cultural pseudospeciation. The fact that the other party is often denied a share in our common humanity shifts the conflict to the interspecific level, and interspecific aggression is generally destructive in the animal kingdom too. Over the biological norm filter that inhibits destructive aggression in man as in other creatures, a cultural norm filter is superimposed that commands us to kill. (I shall discuss the resulting conflict of norms later.) The important point to bear in mind is that destructive war is a result of cultural evolution. Furthermore, it is not, as is sometimes maintained, a pathological phenomenon, but performs important functions, as we shall see later. It also accelerated biological and cultural evolution by the intensification of selective pressures. This applied both to the rapid development of the brain and to the development of altruistic behavior.[2] The question remains open whether humanity can break out of this self-reinforcing process of increasing aggressivity or is bound to go on passively subjecting itself to it.

There are elements in our motivational structure that make development in the direction of peace possible. For instance, in intergroup conflict the inhibitions on aggression are not totally eliminated. Also, selective pressures similar to those at work in the intraspecific aggression of animals seem to be working toward a cultural ritualization of human intergroup aggression. But there are limits to the automatism of this development. In the animal kingdom, conflict is completely ritualized only if the losing party escapes and thus the final situation—"enemy no longer present"—has been attained. But in conflicts between human groups, this is no longer possible, because on our overpopulated planet, little empty space is left to which defeated groups can escape.

Before coming to grips with these special questions of the possible ritualization of war and hence the abandonment of destructive intragroup conflict and the possibilities of further development toward peace, we must examine more closely the phenomenon of war and its dissemination, the forms it takes, and above all its functions.

Clearly, I make a sharp distinction between war as a product of cultural evolution and the essentially biologically determined forms of individualized intragroup aggression. The destructive character of war developed culturally hand in hand with man's pseudospeciation. That does not mean it has no biological roots. We are preadapted to it both by our inborn rejection of strangers and our inborn readiness for aggressive action. The fear of strangers is still an important factor in releasing group aggression, and it is utilized in the collective aggression of war. War developed as a cultural mechanism for spacing out strangers, and in this territorial function, it is comparable with biologically determined forms of territorial aggression. It is wrong to equate it with a pathological degeneration such as murder, as is done by, among others, Fromm, who sees the root of the evil in sadism and necrophilia.[3]

Conradt takes the view that uninhibited intergroup aggression—war—is a species-specific characteristic of man. If he had said universal instead of species-specific, I should have agreed. But he holds that uninhibited intergroup aggression is a congenital human characteristic. He argues that the acceleration of human evolution is explicable only on the basis of intergroup selection, that increasing stress does not cause murder rates to escalate into warlike clashes ("as the environmental theory would lead one to expect"), that war is completely different in kind from escalated criminality, and that wars take place in the absence of any kind of stress situation—all of which is true. He also points out the double morality that exists in relation to killing, which results in the fact that killing in war is not punished even when the victims are defenseless women and children.

He mentions the extraordinary difference in scale between the death rate from murder and the death rate from war, and he claims that the idea of an intergroup aggressivity peculiar to the human species explains "why doctrines of salvation as an expression of superindividual will" have always spread aggressively in spite of the principles they claimed to be propagating. "The religion of love of one's neighbor was propagated with the sword, military dictatorships are imposed on nations in the name of freedom and democracy, and party dictatorships are established to liberate the working class."[4]

Conradt rightly sees the necessity of differentiating between intergroup and intragroup aggression, but his conclusion that war is a new phylogenetic acquisition of man is not convincing. The mechanisms of pseudospeciation that result in groups' marking themselves off from and dehumanizing other groups and the invention of weapons—all of which are preliminary conditions for the destructive aggression of war—are clearly the result of cultural evolution.

In stating this, I hope I have made it clear that I do not regard the phenomenon of war either as the result of an aggressive drive periodically pressing for discharge or as a way of behavior that is constitutional in man. Such views have occasionally been attributed to us ethologists by critics.[5] I have found nothing in Lorenz to justify such a conclusion.* He merely emphasized that man's phylogenetically acquired aggressive disposition can be used in war. It is certain that the decision to make war is often the result of cold calculation and planning by tribal chiefs or statesmen. But it is equally certain that efforts are made to engage the individual combatant's emotions. In this respect, there are links between individual aggression and war.

PARADISE LOST?

The idea of a paradise from which we men were driven reappears in the literature of aggression. A number of authors insist that, once

* In *On Aggression,* he did not, however, distinguish clearly between intragroup and intergroup aggression. In an interview, he said: "If I were to write *On Aggression* again, I would make a much stricter distinction between individual aggressivity within a society and the collective aggressivity of one ethnic group against another. These may well be two different programs. They appear to be different in animals. The behavior patterns of animals seeking status and fighting for rank order are entirely different from the behavior pattern of the whole group fighting another group. I may have been wrong in not distinguishing precisely enough between these two factors" (*Psychology Today* [November 1974]: 89–90).

upon a time, man was a peaceful creature, and that greed for possessions and strife developed only with the development of horticulture and agriculture. From this, they conclude that man is not congenitally intolerant of his fellows. Many seem to find this comforting, for it makes it look as if man is not so bad after all, so things will be all right in the end. In the discussion of basic human nature, Rousseau's theory of the inherently peaceful nature of primitive man has been revived. Reynolds, for instance, states that the facts show that Paleolithic man was not territorial and to a large extent lived in open groups: "The evidence indicates that early paleolithic man was not cooperative, not territorial, and had social and sexual relationships over wide areas (evidence based on Vallois). Societies still living in a nomadic hunter-gatherer ecology, such as the Bushmen of the Kalahari or the Hadza of East Africa, show little territoriality or intergroup aggression."

The "evidence" on which Reynolds relies consists only of Vallois's statement that *present-day* hunters and food gatherers do not seem to be territorial; so far as our ancestors are concerned, he *conjectures* that seasonal climatic changes and the resulting animal migrations during the Wisconsin (fourth glacial) period must have caused migrations of hunting groups, which in his view is not consistent with the existence of definite territories.[6] Furthermore, the great morphological similarity between some Cro-Magnon groups and Neanderthalers seems to point to the existence of direct relations between them in spite of their geographical separation.

These last statements are obviously speculative and do not provide very solid support for Reynolds's argument. That migrations are perfectly compatible with territoriality is shown by the subarctic Indians of Canada,[7] and history right up to the present day shows that territoriality sets up no absolute barriers. There remains the reference to contemporary hunters and food gatherers. But before dealing with the latter, it will be worth lingering a little longer over the archaeological evidence.

Dart observed that many of the Australopithecus skulls he found showed injuries pointing to the use of violence.[8] Roper examined the accumulated material on bone damage to Australopithecus man, men of the Pithecanthropus group, and European fourth-glacial and pre–fourth-glacial men and, applying very critical standards, concluded that a substantial proportion of the injuries were the results of combat.

These findings were confirmed by Mohr, who reported a total of

158 bone injuries from the Paleolithic, Mesolithic, and Neolithic periods. Most of these injuries, 96 altogether, came from the Neolithic. Of these, 62 percent had healed; 47 of the injuries were fractures of the skull, 16 were of the upper extremities, and 14 were of the lower, 16 were of the spine, 3 of the breastbone and 1 of the pelvis. Most of the injuries, according to Mohr, were caused by stone axes. In the bones of the spine and the lower extremities, she found injuries caused by arrowheads; in some cases the wounds had healed with stone arrowheads still in position.[10]

That men killed each other in the Stone Age, and in all other ages as well, is also illustrated in the rock drawings in Western Europe. Pictorial evidence from the Mesolithic era showing men confronting each other in warlike postures is to be found in Kühn, (figure 12).[11] Kühn describes the pictures in the Valltorta ravine (near Albocater, Spain), in a niche of the Saltadora cave:

> I see a picture of a hunter falling to the ground after being struck by an arrow. One leg is stretched forward, his hand rests on his knee. A head adornment, a kind of crown, is falling from his head; his right hand still clasps the bow, but his enemy's arrows have pierced him, his life is over. Thus even in earliest antiquity men killed one another. Where is paradise receding to? Is it a dream of humanity? Is war, fighting, the purpose of human life? Is fighting as old as life itself? Here are primeval pictures of humanity on this earth. Primeval pictures antedating all memory, older than all myth and legend—and already we see the killing of man by man, already we see fighting, already we see war.[12]

Neolithic rock paintings also sometimes show men being killed by their fellows. The first fortified settlements appear at the beginning of the Neolithic era, and many of them show traces of violent destruction. Many axes dating from this period were unsuitable for hunting and were clearly battle-axes.[13] So much for the peacefulness of our Stone Age ancestors.

Warlike competition between human groups played an important part at the dawn of human history, and as Bigelow points out, reluctance to admit this is based not so much on lack of facts as cn refusal to acknowledge them: "Difficulties in assessing the role of intergroup competition in human evolution are due mainly to our very strong reluctance to face the facts, not to a scarcity of evidence. There is overwhelming evidence of man's potential for violence, and this may be precisely why it frightens us."[14]

Sometimes it is maintained that warlike clashes in prehistory were

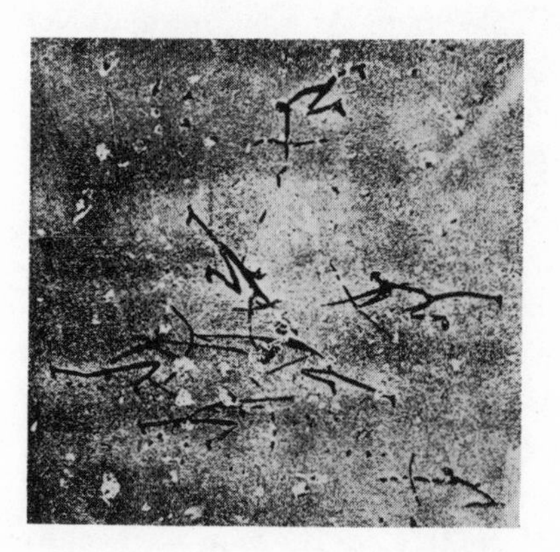

Figure 12. Battle scene, in a paleolithic cave painting at Morela la Vella, Castile, Spain. From Kühn (1929). Copy by Kacher.

Enlargement of the above

unlikely because of the small size of the human population, but that argument has not been properly thought out. We know that hunters and food gatherers need very large territories, and conditions are not the same everywhere. Some areas are rich in game and plant food, firewood, and watering places, and other areas offer less favorable living conditions. There is not the slightest reason to suppose that our ancestors did not compete for the better areas. The archaeological evidence shows that this competition was warlike.

TERRITORIALITY AND AGGRESSIVITY AMONG HUNTERS AND FOOD GATHERERS

In a number of recent publications, "hunting and food-gathering tribes" have been represented as peaceful, lacking in aggression, and not territorial.[15] DeVore puts it poetically: "The Bushmen and the hunter-gatherers generally have what in the modern idiom might be called the 'flower child solution.' You put your goods on your back and you go. You do not have to stay and defend any piece of territory or defend fixed assets."[16]

This is a surprising claim, since a glance at anthropological literature shows that there are a large number of thoroughly warlike tribes of hunters and food gatherers who defend territories. Of the 12 cultures mentioned in Bicchieri's book on present-day hunters and food gatherers, none is specifically described as nonterritorial. Exclusive territoriality is specifically attributed to 4 of them, and possession of territories can be deduced from the description of 5 others. No positive conclusion can be drawn from the facts given about 3 others, but 1 of these 3 are the !Kung Bushmen, who must be regarded as territorial on the basis of other reports.[17] According to Service, all hunting and food-gathering tribes possess territories that are normally closed to strangers.[18]

Anthropological findings provide equally slender justification for attributing any special peacefulness to hunting and food-gathering cultures. Divale studied sexual relationships among 99 hunting and food-gathering bands belonging to 37 different cultures. Of these, 68 bands belonging to 31 cultures still engaged in war at the time of the study; 20 bands belonging to 5 cultures had ceased waging war from 5 to 25 years earlier; and 11 bands from 5 cultures had done so more than 25 years earlier. Some of the cultures were represented in all 3 categories.[19] All the bands studied, therefore, had at least a warlike past. This may suffice to show the absurdity of the claim that most hunters and food gatherers are peaceful. A great deal more information about territoriality among hunters and food gatherers is to be found in Hobhouse and Frobenius.[20]

If one looks at the evidence that forms the basis for the claim that man's nature was originally peaceful, one quickly discovers that the only hunters and food gatherers who can be classed as peaceful are the Eskimos, the Hadza, the Pygmies, and the Bushmen of the Kalahari. Schjelderup makes the surprising statement that the Kwakiutl Indians lack the "fighting instinct,"[21] but that need not be taken seriously, since we know from Boas about the potlatch feasts at which the Kwakiutl chiefs competed in shaming their guests by extravagantly destroying valuable possessions. They actually called these feasts battles, and the songs sung at them were aggressive.[22]

I quote examples of such songs, which I took from Benedict, in my *Love and Hate*. Let me repeat some of them here. A Kwakiutl chieftain boasted to his guests: "Furthermore, such is my pride that I will kill on this fire my copper Dandalayu, which is groaning in my house. You all know how much I paid for it. I bought it for four thousand blankets. Now I will break it in order to vanquish my rival. I will make my house a fighting place for you, my tribe. Be happy, chiefs, this is the first time that so great a potlatch has been given."[23]

Another chieftain sang: "I search among all the invited guests for greatness like mine. I cannot find one chief among the guests. They never return feasts, the orphans, poor people, chiefs of the tribes! They disgrace themselves. I am he who gives these sea otters to the chiefs, the guests, the chiefs of the tribes. I am he who gives canoes to the chiefs, the guests, the chiefs of the tribes."[24]

I find it hard to see in what way aggression is alien to these tribes. Schmidbauer says that the Kwakiutl destroyed only their own property.[25] That is true, but included in that property were slaves, who were killed with special clubs.

Certain tribes that practice agriculture and horticulture are repeatedly held up as models of aggression-free social life. One of these is the Zuñi, of whom Helmuth says: "There is not one sentence in R. Benedict's description of the life of the Zuñi that points to any kind of aggression among them."[25] Weidkuhn points out that he cannot have read Benedict very attentively, because the initiation rituals she described are exceedingly aggressive.[26] As for the allegedly nonwarlike Arapesh, we now know that they make war.[27] But let us remain with the hunting and food-gathering cultures for the moment. What is the evidence for the alleged peacefulness of the Eskimos, Pygmies, Hadzas, and Bushmen? Does it stand up to critical examination?

The Eskimos

Reports of the peacefulness of the Eskimos go back to Fridtjof Nansen, in whose portrayal they did not kill one another or make war. But König pointed out that these reports were colored by the wish to show the Eskimos in a favorable light: "Nansen saw very little of the Eskimos in their natural state, and his moral judgment on them, especially in his *Eskimo Life,* is thoroughly tendentiously colored, as he wished to arouse sympathy."[28]

In fact, we know of many more or less ritualized forms of aggression among the Eskimos, ranging from wrestling and fist fights to the celebrated song duels of the West and East Greenlanders.* Also Eskimo stories and soapstone carvings are concerned with acts of violence and murder (figure 13). Territoriality is by no means unknown among them. Petersen gives an account of the hunting territories of the West Greenlanders (Sukkertoppen District), which surrounded the group camp. The Eskimos and their neighbors, he states, overstepped hunting territory boundaries only at their peril; the penalty might be death.[29] This applied not only to relations between Eskimos and Indians, but also to those between different Eskimo groups, for instance, the Akudnirmiut and Tununermiut,[30] the Cape York and northwest Greenlanders,[31] and the Copper and Netsilik Eskimos.[32]

* I pointed out in 1970 that the Eskimos cannot be regarded as especially peaceful. I mentioned the various levels of ritualization of dueling (fist fights and song duels), referred to König's statements about Nansen, and finally also mentioned the incidence of family quarrels. In a reply, Plack states: "The Eskimo men whom Eibl-Eibesfeldt portrays as violent primitives who beat their wives are no longer those whom Nansen described at the time when they were beginning to be converted to Christianity; at that time, they were still shocked at how white sailors quarreled and fought. Eibl relies on Rasmussen, who described a brawl between one married Greenland couple—an individual deviation into aggression such as presumably may occur anywhere." (*Der Mythos von Aggressionstrieb* [Munich, 1973], p. 34.)

Those who quote in this way are not seriously interested in establishing the truth.

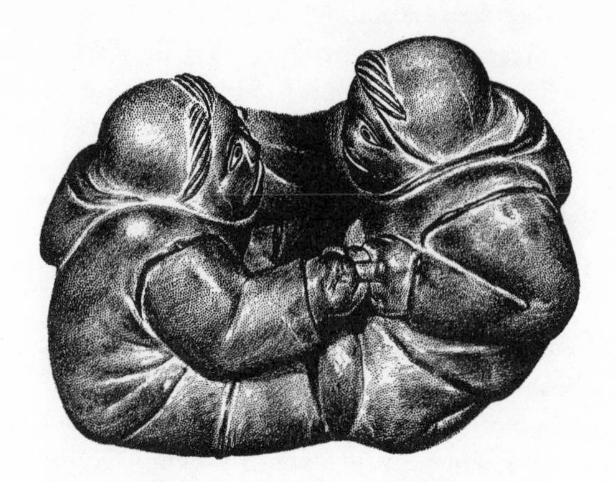

Figure 13. Two Eskimo soapstone carvings from Povungnituk, Hudson Bay. They show a finger-tugging contest and a duel with knives. Drawings by H. Kacher from photographs in Nungak and Arima (1969).

Petersen says that murderers left their groups and went to live outside the settlements,

> as their killing had created enemies for them. Their habitations were chosen more for reasons of safety than for the number of animals available. If they discovered a place where there was but little chance of being surprised, they remained there permanently, even though the wild life of the district might not stand much thinning out. Having become settled, force of habit began to assert itself: they regarded the place and the district as their own, and other hunters they met as invading competitors, apart from the fact that they might be enemies. This led to fresh murders and new enemies who also had to be disposed of.
>
> Tusilartoq first killed a man because he had ventured into the hunting territory of the Umanaq people (at Isortoq), and there met a man who attempted to kill him (Lynge 1939, pp. 66–67; Rasmussen 1924, p. 317). Sturdy Qagssuk had a dwelling place to himself, but was a peaceful man. His son was attacked by his brother-in-law, and only escaped at the last moment. This infuriated Qagssuk, who exterminated the whole settlement. Thereafter he would not tolerate strangers in his area, but curiously enough did them no harm if they said they had lost their way.[33]

Petersen also says that certain families had definite places for netting seals, which they kept as their hunting territory for generations, and the same applied to trout-fishing places. An orderly division of territories for reindeer hunting seems also to have existed. Petersen describes how an Eskimo instructed his son in reindeer hunting: "On no account must you hunt in an easterly direction, for there Serquilisaq has his camp. He killed your elder brother, just when he was beginning to become a good hunter."[34]

There were hunting and fishing grounds that belonged to families, and others that belonged to the group collectively. Petersen comments that he was surprised to find that Eskimos had "private territories," that is, family rights over certain localities. He also observes that there was a conflict between family privilege depending on priority and the right of a group to hunt everywhere, but family rights were respected and were nullified only by lack of use.

Finally, he describes the difficulty of finding out details about conditions among other Eskimos in Greenland:

> My question as to whether some families had fixed summer camps nearly always elicited the reply that everybody knew that all were allowed to hunt anywhere. This answer is true enough, but further

questions nevertheless showed a survival of fixed place rights and the right of use connected with it. From Northern Greenland there were a few who could remember that net places were reserved to their users as long as they occupied the site, but their right lapsed when they began to hang up nets by icebergs. South of Sukkertoppen I have not had the question confirmed and from Angmagssalik only received one uncertain confirmation. This suggests that it is something that is in the process of being forgotten, and that earlier it was more widespread.[35]

The older literature on the subject contains continual references to territorial division among the Eskimos. Klutschak writes:

There is no basis for regarding the Eskimos as nomads, as they are traditionally tied to settlements in certain reservations from generation to generation and can cross their boundaries only with the consent of their neighbors. Only within their own hunting territories do they change their place of residence with the change of seasons and the concomitant change in the country's game.[36]

Klutschak also mentions that many tribes had suffered severely in warfare. The Ukusiksilik Eskimos were the survivors of what had once been a big tribe whose home not very long before had been on the west coast of the Adelaide Peninsula. They had been greatly weakened by prolonged warfare with the Ugzulik and Netsilik tribes that were now settled there, and had been forced to leave their old hunting grounds. When Klutschak visited them, the whole tribe had been reduced to only sixteen families. He writes concerning their blood vengeance:

For a long time, there had been a feud between the Netsilik and Eivilik Eskimos. Its origin was to be sought among generations long since dead, and it has been continued only as a result of the blood vengeance that still generally prevails among the Eskimos. It was this blood vengeance that kept our Eskimo Joe in a continual state of anxiety when he was near the Netsiliks. . . . After further discussion it was decided that the reconciliation should take place at a general meeting. . . . The individuals, all men, arrived armed with knives, and the negotiations, like everything else, began with a light meal. . . . The meal was followed by a long discussion, and not until about two hours had elapsed did everyone lay aside his knife, and the two formerly hostile parties laid their hands on each other's breast and spoke the word *ilaga* ("let us be friends"). The two sides parted in evident satisfaction, and a joint entertainment was arranged for the evening.[37]

In another instance, a Kinepetu Eskimo who was a guest of the

> Eivilik Eskimos was slightly wounded at target shooting. His relatives fetched him and demanded compensation, which was refused. Thereupon, the claimants chose three men who, as representatives of the whole tribe, declared a feud against three male members of the Eivilik tribe in case their demand was not met. The two tribes would go on living in peace with each other, but each of these six individuals would cross the borders of the other side's hunting grounds only at the risk of being killed. . . . Such minor incidents often give rise to long-standing feuds that continue for generations in the form of blood vengeance. . . . How scrupulously the duty of exercising blood vengeance is carried out was shown by the example of our Eskimo Ikuma, who undertook a 400-mile journey in the depth of winter to carry out this duty against the murderer of his uncle, a Netsilik Eskimo.[38]

The fact that in this instance a few selected men acted for a whole community is an interesting form of ritualization that limits bloodshed, though it does not prevent it. Among the East Greenlanders, ritualization goes so far that even murder can be compensated for in a song duel[39]—quarrels about women are a frequent reason for murder and manslaughter.[40]

When the Bering Eskimos wanted to make war, they first sent messengers to friendly groups to inform them of their intentions. They would then surreptitiously surround the enemy village, creep up to the houses at night, barricade the entrances from outside, and shoot those inside through the smoke holes. They would then loot the houses and the bodies of the dead. Bristol Bay Eskimos took away heads as trophies, put them on poles, and stuck arrows crosswise through the noses. This was the general practice among these Eskimos, according to Nelson. He also reports that hostile groups of Eskimos showered each other with arrows in open battle, and that often men looked like pincushions before they fell. If one side was tired or hungry, a fur jacket was raised on a pole. This was an invitation to a truce. If the other side agreed, there would then be a pause in the battle. A few guards remained on duty to ensure that the truce was observed. When it was over, the battle was resumed.[41]

It is evident that the Eskimos can hardly be regarded as especially peaceful hunters and food gatherers. Knowledge of their aggressivity is by no means new, but some believers in the neo-Rousseauist mythology obviously are unwilling to accept it.

If most Eskimos—with the exception of some groups in Alaska—now live in open, nonterritorial communities, that is the result of contact with civilization, which has led to a reduction in population, migration, and the breakup of old group structures. Boas reports that with the arrival of whalers in Baffin Bay, many Eskimo families left their homes, as they were attracted by the white man's goods and started bartering with him. Syphilis, diphtheria, and pneumonia spread as a result of the contact with Europeans and killed many Eskimos. When white men arrived in Cumberland Sound in about 1840, the Eskimos population there consisted of about 1,500 souls; by 1857, there were 300. The consequence of such a drastic reduction in population is known to be that survivors of different groups amalgamate with each other. According to Boas, the Oqomiut had previously been divided into four subtribes known by the name of their territories: "their old settlements are still inhabited, but their separate tribal identity is gone, a fact which is due as well to the diminution in their numbers as to the influence of whalers visiting them."[42]

Service gives further examples of this, and concludes: "It seems clearly evident that aboriginal Eskimo society was not fluid, informal and composite, nor was it a family level of integration caused by the nature of the game hunted. The later composite groups of unrelated Eskimos known to modern ethnology are readily explained as a consequence of direct and indirect European influences that can only be described as catastrophic."[43]

The Pygmies

The view that Pygmies are peaceful hunters and food gatherers goes back to Turnbull's studies of 1961 and 1965.[44] In the latter monograph, however, he is more cautious than those who quote him, for he points out that very little is known about the relations between different bands. In his contribution to Lee and DeVore's symposium in 1968, he emphasizes the Pygmies' peacefulness—"there is an almost total lack of aggression, emotional or physical, and this is borne out by the lack of warfare, feuding, witchcraft, and sorcery"[45] —only to state shortly afterward that there is a great deal of strife within the horde, particularly between married couples.

If we turn to the older literature, however, we find reports of war, blood vengeance, and territoriality. Schebesta says that the Bambuti Pygmies of Ituri have forest territories,[46] and he later mentions roving and hunting territories that are regarded as group property.

The roving and hunting territory of the Bambuti group, which is marked off by precisely known boundaries, belongs to a definite group of related families who alone have a claim to it. They alone, to the exclusion of all other groups, are allowed regularly to seek their food within this territory. All the members of the group enjoy the same unrestricted right in this respect. The intrusion of strangers for the purpose of hunting or gathering food is not permitted and leads to quarrels and wars. Friendly neighboring groups with whom marriage ties exist are, however, sometimes allowed the right to cross the boundaries. My observations repeatedly showed me that the Bambuti greatly dislike crossing into strange territory and went there only on my insistence or at the bidding of my host. Pygmies are doubly shy and nervous on foreign territory. . . . Groups settled in foreign territory—as was frequently the case during my last journey to the big camps—always hunt in their own hunting territory.[47]

On the subject of nomadism, Schebesta writes: "Nomadism does not mean aimless roaming, but traveling to a distant area with limits established in relation to neighboring groups. I should like to place special emphasis on this, as unrestricted nomadism is sometimes attributed to the Bambuti. Their travels in the forest are neither purposeless nor aimless."[48]

Bicchieri also states specifically that the Bambuti Pygmies have precisely demarcated territories:

Concomitant to the abundance and uniformity of the distribution of resources throughout the vast forest and throughout the year, the Bambuti band is not "forced" into a specific area. Yet both net hunters and archers are very definitely associated to specific territories. It could be suggested, furthermore, that a specific territory, delineated by natural boundaries, "owns" its band (see Bicchieri, 1965). Three conditions, (1) favorable people—land size ratio, (2) natural boundaries, and (3) plentiful and uniformly distributed resources, led to the presence of discrete territorial areas. A specific band "belongs" to these *discrete* areas and acts as if it had the resources of the particular territory in stewardship.[48]

Bicchieri also says that the hordes are known by the name of their territories and that their "nomadism" really consists of cyclical migrations within their territory.

Because of the temporary nature of the camp and the easiness with which it can be set up at any point in the home territory, the Bambuti think of themselves as having a stable pattern of residence, more stable, in fact, than that of the people of the neighboring village of the Negro agriculturist, who, because of soil exhaustion and forest encroachment, moves his "stable" residence every few years.[49]

Schebesta also describes Pygmy warfare, which consists primarily of attacking from ambush:

Just as the Pygmy hunter stalks and outwits his prey, so does he stalk his enemy, lie in wait for him, and shoot him down with an arrow from safe cover. An elder of the Mamvu-Efe gave me a vivid account of such an incident. When he was young, he told me, his clan made its way into a strange area beyond Mambasa. He was still a boy at the time, and was cracking nuts with his brother. His father began to feel uneasy and warned them and told them to come away, and they promised to follow him as soon as they had finished their work. A few minutes later, a Mombuti appeared in the bush. The boy sprang to his feet, but an arrow was already in his side; he showed me the scar. He collapsed, his brother took him by the arm, but an arrow struck him too. The boys' cries caused their father and his people to hurry to the spot. Four men were apparently killed in the battle that ensued, and the bodies of three of them were left lying there.

The occasion for this bloodshed was a Mamvu-Efe incursion into strange territory; it was an infraction of the sovereignty of another clan. This was generally recognized by the Bambuti as a legitimate reason for hostilities among themselves. For the same reason, the Bambuti fought the Negroes who advanced into the forest and ultimately also the white man's caravans.

The Bambuti method of warfare differs from that of the Negroes. They never advance in a body as the latter do, but scatter as they do when hunting, creep up to the enemy through the thick undergrowth, shoot their poisoned arrows, and vanish again as noiselessly as they came, without the enemy's being able to tell where the arrows came from or how many assailants there were. Alternatively, the Bambuti will lie on the ground, shamming death, and allow the enemy to draw close to them, whereupon they will shoot a few arrows at them and again quickly vanish into the undergrowth. This way of fighting should not be interpreted as Pygmy cowardice, for it is in accordance with their nature as hunters in the primeval forest. They use the same method in private feuds or in acts of blood vengeance and dispose of their enemies or victims from an ambush.[50]

Schumacher says that the war weapons of the Kivu Pygmies consist of lances, sickle-shaped knives, bamboo bows and arrows, clubs, and a big, broad, woven shield.[51] The shield provides conclusive evidence of adaptation to intraspecific conflict. There is no use for a shield in hunting.

In view of all this, Pygmies can hardly be described as being especially peaceful, nor as nonterritorial.*

The Hadza

The Hadza of Tanzania have recently been quoted as providing further evidence of the basic peacefulness of hunters and food gatherers. Woodburn reports that they have no territories, show no aggression, and live in open groups.[52] He studied them in 1958 and the following years. But the 2,000 square miles in which they originally lived had by that time been reduced to only a little more than 750 square miles, and when uprooting takes place on such a scale, changes in behavior and social structure must be expected.

Kohl-Larsen, who studied the Hadza in 1934, 1936, 1937, 1938, and 1939, paints a picture of them that differs from Woodburn's in many respects. First of all, he says:

> The individual Tindiga [another tribal name of the Hadza] hordes do not roam all over the country, but each of the three keeps to a quite definite area. I never saw the Matete horde hunt in the hunting area of another horde. The northeastern boundary of their hunting area is the Dumungiddah, and in the east, it is the last foothills of the Iraku mountains, which they call the Kidabimbirigaah. To the south, they never extend their hunting expeditions beyond the mountains that separate the Ngara Lake from the Hohenlohe Rift. On the other hand, I was assured that when Shungwitcha, the leader of the Matete horde, went to see his brother on the "Great Water," he was allowed to hunt there. And in times of famine, the boundaries of the individual hordes are lifted, so that they are allowed to hunt in areas allotted to other hordes.† Quite apart from the fact that active traffic between different hordes does not exist, it seems to be the usual hunting practice that each horde respects its neighbor's hunting boundaries.[53]

Kohl-Larsen also shows that the Hadza used to engage in war-

* See also the recent publication of M. Godelier, "Territory and Property in Primitive Society," *Social Science Information* 17 (London, 1978): 399–426.

† Compare the description of the Bushmen's nexus system on p. 149—I. E.-E.

‡ Bagshawe, who visited them in 1917, also mentioned that feuds with fatal outcome were by no means rare. "I do not mean to infer that the Kangeju [another name for the Hadza] are entirely law-abiding, for I am well aware that numerous deaths have occurred as a result of feuds between individuals and families and between them and the Dorroggo" ("The Peoples of the Happy Valley [East Africa]," *Journal of African Society* 24 [1924/25]: 123).

fare.‡ He quotes an account by a Hadza informant, dated July 8, 1938:

> In the old days, the Hadzapi made war on each other. A band that lived in Mangola went to another band over in Lubiro. When it arrived there, a man was chosen who went to the Lubiro band and said: "We have come to you today to make war on you." One of the Lubiro horde said: "Well, if you have come to make war, that is satisfactory to us." The people in Lubiro met together and consulted. They chose a man to fight the man from Mangola. When he had been chosen, each of the two was given two sticks with which they fought each other. If neither of the two was beaten, the two bands began fighting each other with arrows and spears. While they were fighting thus, an old woman advanced from one band and an old man from the other. They took up a position between the two fighting bands and said: "Sit down all of you, and rest a little." After they had rested for a little while, they again began fighting. They fought each other for a very long time. When a horde is beaten, it runs away. The other one, the winner, chases it for a time and then goes back to its camp and sleeps. Next day the horde that won goes to the losing one and spends the night with it. In the morning, all the strong men and youths gather together and go hunting. When they have killed a few animals, they take the meat that is fat and sit down to eat. Only men may be present. If a woman joins the men, she may be killed. But if the woman is carrying a child on her back, she pinches it and makes it cry. If the men hear the child crying, they cannot strike the woman. When they have eaten meat together, they are friends again. If they live like that for several days and see they are living for nothing [have no work and nothing to do], they seek out another band and make war on it. They fight so much that often several men are killed. Those who are beaten go over into the great host of the winners.[54]

This report is notable not only for its many details, but also because it shows the fallaciousness of concluding from the present behavior of the Hadza that they lived peacefully in the past.

The Bushmen

TERRITORIALITY

Bushmen play a big part in the discussion of aggression as a model, since it is claimed that theirs is a model peaceful culture. As I had the good fortune to pay five visits to the Central Kalahari and South-West Africa, enabling me to make a thorough study of the !Ko, G/wi, and !Kung Bushmen, I propose to give a rather detailed account of their aggressive behavior. I discussed them in a monograph,[55] but have since gathered additional material. Also a number of remarkable new publications have appeared with which I propose to deal.*

Sahlins[56] and Lee say that the Bushmen have no territories and live in open societies. Lee writes: "The camp [of the !Kung Bushmen] is an open aggregate of persons, which changes in size and composition from day to day. Therefore, I have avoided the term 'band' in describing the !Kung Bushmen living groups. Each water hole has a hinterland lying within a six mile radius which is regularly exploited for vegetables and animal food. These areas are not territories in a zoological sense, since they are not defended against outsiders.[57]

This definition of territory is based on the mistaken idea that territoriality is always expressed in fighting. As I explained on page 42, that is not the case. Territorial integrity can also be preserved by ritualized forms of demonstrating possession, as is pointed out by, among others, Rappaport, Ortiz, and Wilmsen.[58] Also the existence of territorial boundaries does not mean that there is no flow of individuals across them. The only restriction is that the owner is dominant in his territory, enjoys privileges in it, and can grant or refuse entry to it.

Marshall describes the Bushmen as peaceful and harmless people, but in doing so, she is merely recording a general impression; on the other hand, she reports privileges enjoyed by certain families in some food-gathering areas.[59]

In the secondary and tertiary literature, these statements are strained beyond all measure to support the theory of man's inherent peacefulness. An example of this is provided by Schmidbauer,[60] who has never in his life had even a distant view of a hunter and food gatherer and makes statements about their peaceful nature all the

* Illustrations of behavior patterns of aggression, aggression control, and group bonding appeared in my monograph; consequently, I shall not reproduce them here. My work is supported by films of unposed interactions. The films were published as part of the Human Ethology Film Library of the Max Planck Society. On the Bushmen project, which will continue for several more years, I am working with the anthropologists Dr. H.-J. Heinz and Dr. Polly Wiessner and my colleagues H. Sbrzesny and D. Heunemann. I wish to thank all four for their harmonious cooperation.

more unrestrainedly for that reason. For information about the Bushmen he relies principally on Lee's older publications, obviously without knowing that Lee has long since revised his views. It is clear from his descriptions that Bushmen own territories on a group basis.[61]

It is surprising that anyone should attribute nonterritoriality to Bushmen, as observers have given ample examples of their territorial behavior, apart from the fact that there is an abundance of Bushman paintings showing battles between groups of Bushmen, between Bushmen and Bantus, and between Bushmen and Hottentots. In interethnic clashes, Bushmen trying to steal the cattle of herdsmen tribes were often the aggressors (figures 14, 15, and 16).

Many of the publications giving information about Bushman territoriality have been in German. Passarge described the !Kung Bushmen as warlike, and said that not only the bands but each family possessed its own collecting grounds. "The division of Bushmen into families has long been known . . . but I have never yet seen it stated that land is the legally alloted property of the family. It is a point of tremendous importance, however, for it is only by taking it into account that one can gain a clear insight into the social organization of the Bushmen."[62]

Zastrow and Vedder report that !Kung Bushmen are never permitted to hunt or gather food in the territory of another group.

> Where the Bushman country has not yet been divided up into farms, but clan territory borders on clan territory, every Bushman knows that he may not hunt or gather food in strange territory. If anyone is found doing so, his life is forfeit. That does not mean that he is necessarily killed. The blood vengeance . . . may be prevented, and the death penalty not carried out. It depends on how angry the minions of the law are and on the degree of fear in which the offending clan is held. This enables us to see the explanation of many Bushman excesses against the property and lives of white farmers. If the strange farmer possesses the watering place and also forbids hunting, a state of war automatically exists.[63] *

* Schmidbauer uses this quotation to try to show that my quotations were taken from the older literature, permeated with the white man's "land-grabbing ideology," which showed not so much the original territoriality of the Bushmen as the arrogance of the colonizers. After repeating this quotation and ignoring others, he writes: "Have the human ethologists here been taken in by the 'aggressive savage' stereotype? The use of the terms 'clan territory' and 'minions of the law' plainly shows that we are

Figure 14. Bushman rock painting from South Africa, showing a fight between Bushmen and Basutos. The Bushmen have put arrows into their headbands like decorations, to have them handy. The Basutos carry shields and clubs. From Bleek (1930).

Lebzelter says that the !Kung are very mistrustful when they meet members of a strange band: "Every armed man whom they meet is assumed to be an enemy. A Bushman may enter the territory of a strange tribe only unarmed. Even at the edge of the farm zone, mutual mistrust is so great that a Bushman sent as a messenger to a farm situated in the area of another clan will not dare to leave the road, which serves as a kind of neutral zone. When two armed Bushmen approach each other, they first put down their arms."[64]

Vedder writes: "Every Bushman clan possesses a clan territory inherited from their fathers. Many clans actually possess two—a summer veld and a winter veld. These territories have quite definite boundaries. A Bushman who engages in hunting or food gathering

---

faced here, not with information gained in the field, but rather with atrocity stories about 'brutal savages' of the kind that have always been told in the wake of colonizers' land grabbing in order to justify their own brutal aggression" (*Die sogenannte Aggression* [Hamburg, 1972], p. 109). Schmidbauer, who likes to be recognized as a spokesman for the anthropologists, received a long-overdue riposte from an anthropologist: His statements that "numerous anthropologists have found that the 'most primitive' cultures have frequently also been the most peaceful," and that "the relative lack of aggression of many hunting and food-gathering tribes . . . belongs as before to the most securely established discoveries of cultural anthropology" were answered as follows by Schindler: "Every ethnologist will note this with incredulous surprise. Schmidbauer here presents himself as the spokesman of a discipline and proclaims discoveries of which the specialists in the subject know nothing. He obviously took the silence of ethnologists about his previous statements for consent or even applause. It never seems to have occurred to him that ethnologists have not discussed his works because they have not regarded them as relevant. The inadequate public-relations work of German-language ethnologists may be regarded as deplorable; nevertheless it would be asking too much of them to expect them to attack every loudly proclaimed statement showing ignorance of ethnological facts" ("Territorialität und Aggression: Eine Erwiderung," *Anthropos* 69 [1974], 275).

Figures 15a and b. Bushmen fighting each other. Bushman rock painting from South Africa. From Bleek (1930).

Figure 16. Bushman rock painting showing cattle raid. The Bushmen are seen driving the cattle away while a rearguard keeps the pursuing Bantus at bay. After R. Andree from Weule (1916).

in the territory of a strange clan can be certain that one day he will be struck by a poisoned arrow."[65] The same author says elsewhere: "There are still enemy clans among the Bushmen tribes, and no one dares to cross the clan boundaries." In the old days, he states, the Bushman clans decimated each other. And at that time, they made piles of stones that looked like ancient graves to mark boundaries in certain places.[66]

Trenk reported that the Namib Bushmen regarded certain water holes and hunting areas as family property. No one was allowed to hunt there without the owner's permission. "Every family has its definite place and territory in summer in the Namib, and as soon as the water or the narras and tsama fruit is finished, in the mountains. . . . If a family's water hole is empty, it may live at another water hole, paying rent. The same applies to hunting if the game has left a family area; rent, as it were, must be paid in the form of part of the game that is caught."[67]

Brownlee and Wilhelm[68] both describe territoriality, the latter giving accounts of battles between various !Kung tribes. He says that the Bushmen's skill in warfare, which earlier—when whole tribes lived in close cohesion—had been at a rather high level, had rapidly declined in the past hundred years. They would never again fight Bantu tribes, but quarrels occurred between different !Kung tribes. Wilhelm writes: "Suppose that after the rainy season, the Karakuwisa, for instance, come a long way down the Omuramba Uamatako in search of bush food and, above all, honey. If these bands on such an occasion come across others belonging to the Otjituo tribe [the !Kung who in Wilhelm's time lived near Otjituo], the result is bitter fighting. If the Karakuwisa go eastward during the rainy season into the territory of the Kaokoveld, there are violent clashes there. Also individual clans feud with each other." Armed men would attack the enemy's wharf at dawn. They would kill everyone who failed to escape, including women and children, set fire to the huts, and take away everything they could carry.[69]

Marshall describes the fear of other !Kung Bushmen shown by the !Kung of the Nyae Nyae area. This was so pronounced that they practically never left this area, because on strange territory, they could count neither on bartering for food with the tribesmen settled there nor on securing permission to gather food. The stranger = enemy equation was also very pronounced among them. The Nyae Nyae !Kung described themselves as "pure" or "perfect" (*ju/oassi*), in contrast to other !Kung, whom they regard as "alien," "dangerous" (*ju dole*), and actually murderous with the deadly medicines they possessed.[70] In another work, the same author says of territoriality: "The !Kung say that one cannot eat the ground itself, so it does not matter to whom it belongs. It is these patches of 'veldkos' that are clearly and jealously owned and the territories are shaped in a general way around these patches . . . the strange concept of ownership of 'veldkos' by the band operates almost like a taboo. No external force is established to prevent one band from encroaching on another's 'veldkos' or to prevent individuals from raiding 'veldkos' patches to which they have no right. This is just not done."[71]

Tobias says that Bushmen move about only within their territory: "Territoriality applies among bands of the same tribe and between different tribes. Intertribal bounds are sometimes reinforced by social attitudes such as the traditional enmity between the Auen and Naron. Under special conditions such as an abundance of food these bounds and the accompanying enmity are forgotten."[72]

Silberbauer found that territoriality was typical of the G/wi of the central Kalahari (figure 17). He uses the ethological definition of territoriality found in Willis (see p. 42), explaining: "Willis' description aptly describes the relationships between G/wi bands with regard to their territories; a visiting band or a single visitor submits to the dominance of the host band either by waiting for an invitation or by seeking permission to enter and occupy the territory."[73]

Silberbauer says that visitors passing through a band's territory on the way to another area, as well as those paying a visit to the band itself, come to the camp and ask permission "to stay in your country and drink your water." This is a standard phrase, used in spite of the fact that for most of the year the water holes are dry, particularly at the time when visits are made. He also mentions that "in each band there are individuals known as *!u:ma* (owner) or *!u:sa* in the case of a woman. In the vernacular account the *!u:ma* is the original founder of a band or his male or female descendant."[74]

After visitors have formally asked for permission, they can stay with the band and can ultimately actually be adopted by it. But this, according to Silberbauer, occurs relatively seldom. It is therefore surprising that in spite of this Silberbauer should describe the band as an open group. He seems to interpret the ethological term "closed group" as implying complete exclusivity. Groups as closed as that are exceptional, however. We call a group closed if there are restrictions on group exchange, in contrast to open groups, in which there are no obstacles to changes in their composition. Examples of open groups are the pelagic shoals of fish, which can be joined at any time by fish of the same species. Baboon, macaque, and langur groups are described as closed, though it is perfectly possible for individual members to change groups.

Lee has now published a much more detailed study of the spatial organization of the !Kung and shown that a genuine territorial distribution exists. He too now states that the water holes are in the possession of individuals described as "owners" (*k"ausi*). Each water hole is surrounded by the territory of the "band" (*n!ore*) from which the group draws its food. An individual can inherit his *n!ore* from his father or mother, or sometimes from both; Lee demonstrates a clear patrilineal tendency among males. Those who have relatives in a camp may gather food in the area. He also found that the area from which a group draws its food is limited, but that this area is not defended.[75] If this is correct—before accepting it, one would first need evidence of an unpenalized infraction of ter-

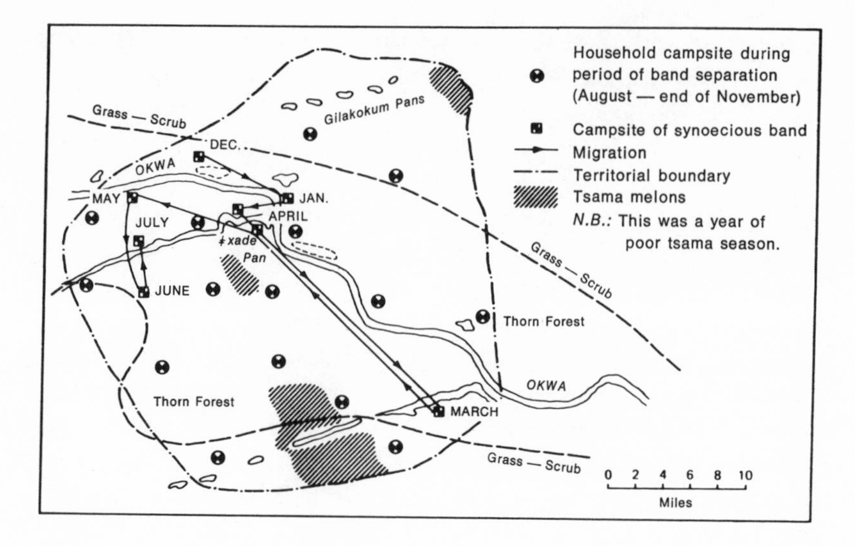

Figure 17. The movement of a G/wi horde over a period of two years. From Silberbauer (1972).

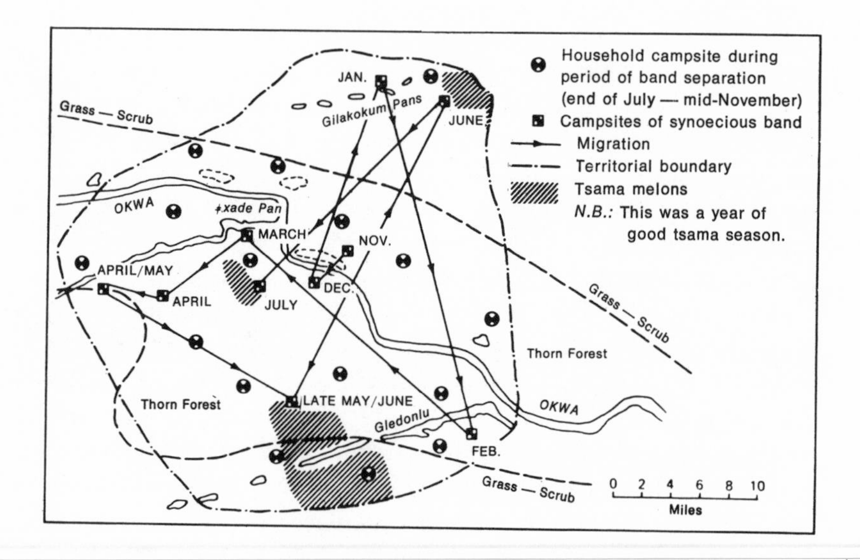

ritorial rules, and Lee gives no examples of this—it might well be a result of the acculturation that has been setting in. In this process unacculturated tribes gradually come under the control of state organs that settle disputes and prevent open aggression.

My own work has so far been concentrated on the !Ko Bushmen of the Central Kalahari, many of whom still live as hunters and food gatherers. I have gathered comparative data among the G/wi and the !Kung.

Heinz notes three levels of social organization among the !Ko, consisting of (1) the family and extended family, (2) the band, and (3) the band nexus. All these units are marked by a clearly defined pattern of bonding and spatial distance keeping. The seating order of members of a family around the fire is less marked than it is among the !Kung,[76] but though a wife is allowed to sit anywhere, her real place is on her husband's right. Parents are expected to build their hut at least 40 feet away from their married children and to ensure that they cannot see them sleeping through the entrance. Every member of a band can hunt and gather food throughout the band territory, but there are areas that are recognized as family areas. A solitary hunter is expected to hunt on the side of the territory adjoining his hut. The same is expected of women gathering food or firewood. This rule is lifted for collective activities.[77]

Bands periodically split up into family groups and move to quite definite family areas, which are respected as such by others. While the family has no direct territorial claim, the band has a clearly defined territory, which is controlled by a headman acting for the group.

A number of bands are associated in a nexus system. The members of a nexus regard each other as "our people"; they exchange marriage partners and share some peculiarities of dialect. Ties of friendship and kinship keep the nexus in being. Its members meet for certain rituals, such as trance dances. In case of need, members of a band can ask to be allowed to hunt and gather food in the territory of another band within the system, and such requests are generally granted. No one would expect such a request to be granted outside the system. The nexus territory is exclusive. I regard this finding of Heinz's as extremely important, for it seems to explain a number of inconsistencies in previous observations, particularly in relation to group exclusivity (figure 18). Members of different bands within a nexus visit each other with relative freedom, which may create the impression of open groups. Silberbauer also found a nexus system among the G/wi. He talks of band alliances. Wiessner has recently demonstrated the existence of a nexus among the !Kung on the basis of a computer analysis of all the observed social contacts.*

Access to a territory is acquired by birth, adoption into a band, or marriage. If the parents come from different bands, married couples have access to two territories. The bridegroom first lives for a time in the bride's territory and thereby acquires access to it. When the couple finally settles with the husband's band, the wife gains access to its territory. This double privilege passes to the children.

My documentation of unposed social interactions (all together nearly 40,000 feet of film about the !Ko are available, as well as 6,500 feet about the !Kung and 45,000 feet about the G/wi) shows that (1) aggressive interactions within the band are frequent; (2) many of the behavior patterns observed in such interactions resemble those that occur in similar circumstances in other cultures; and (3) the situations in which aggressive behavior appears also resemble those in other cultures. The most important of these are are the following:

Sibling Rivalry. Advocates of frustration-free education are fond of referring to the allegedly happy, frustration-free development of children among primitive tribes, no doubt because they have no real knowledge of primitive cultures. The birth of a baby in a Bushman family leads to intense jealousy on the part of an older child. The !Ko accept this as an inescapable fact of life. Rivalry is not limited to sex. Girls are just as jealous of a newborn brother's bond with the mother as they are of a sister's, and older boys compete with a newborn baby for the mother's favor in exactly the same way. As soon as the new sibling is eight to ten months old, it energetically defends its position against its still jealous siblings.

* She found three such nexuses among the northwestern Botswanas. Each consisted of about 100 to 300 individuals. Their kinship ties permitted the members of a nexus generally to move about freely in a nexus area. The nexuses found included (1) the Nxau Nxau, Tsodilo, /gada, and Chenepu area; (2) the Dobe, !xabe, Mahopu, Bate, !ubi, !gose, !angwa, and /ai /ai area; (3) the Ghanzi area. Internexus contacts and marriages also take place, but are much rarer and much less open (all information in a private communication from Polly Wiessner, who generously put these data at my disposal).

I filmed a particularly dramatic case of sibling rivalry among the !Kung. A boy of about four was competing for his mother's favor with his brother, aged about one. The older boy sought contact with his mother, who did not respond, though he used the most varied appeals for attention, offering himself for delousing, holding out the soles of his feet as if he had stepped on a thorn, and several other actions.

He tried pushing, scratching, driving away, and striking his

Figure 18a. Map of Botswana. The hatched portion is enlarged on the following page.

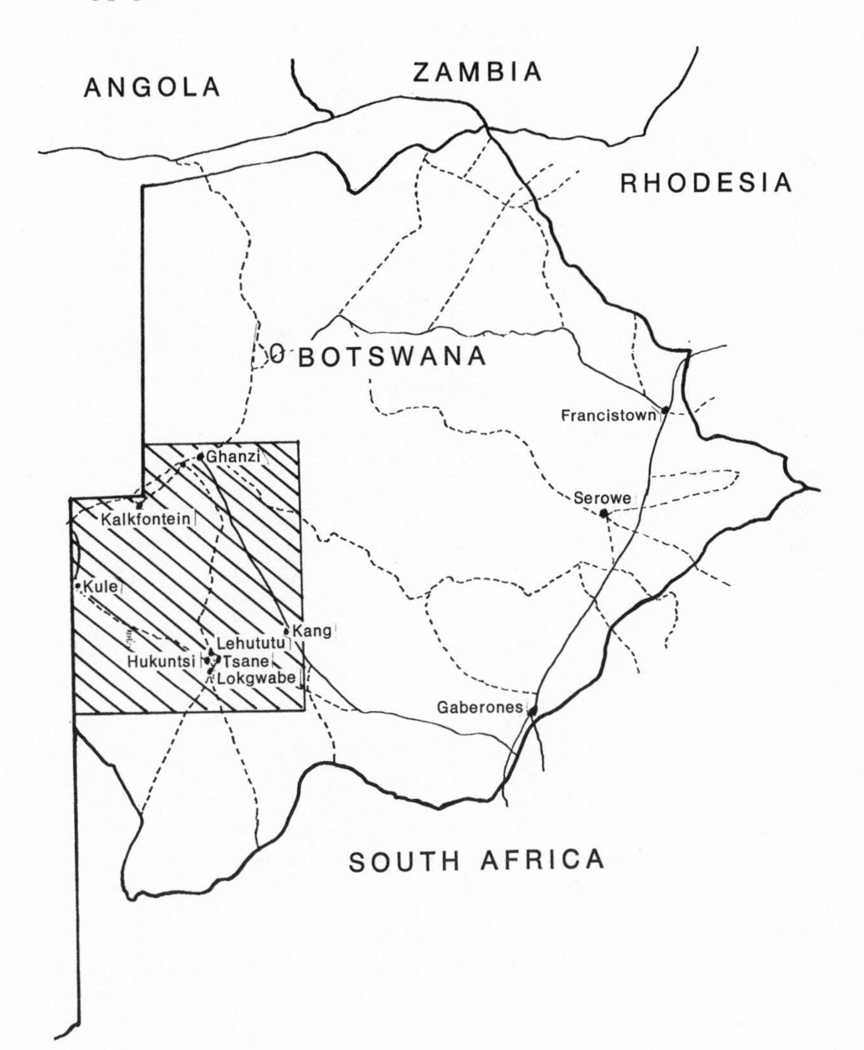

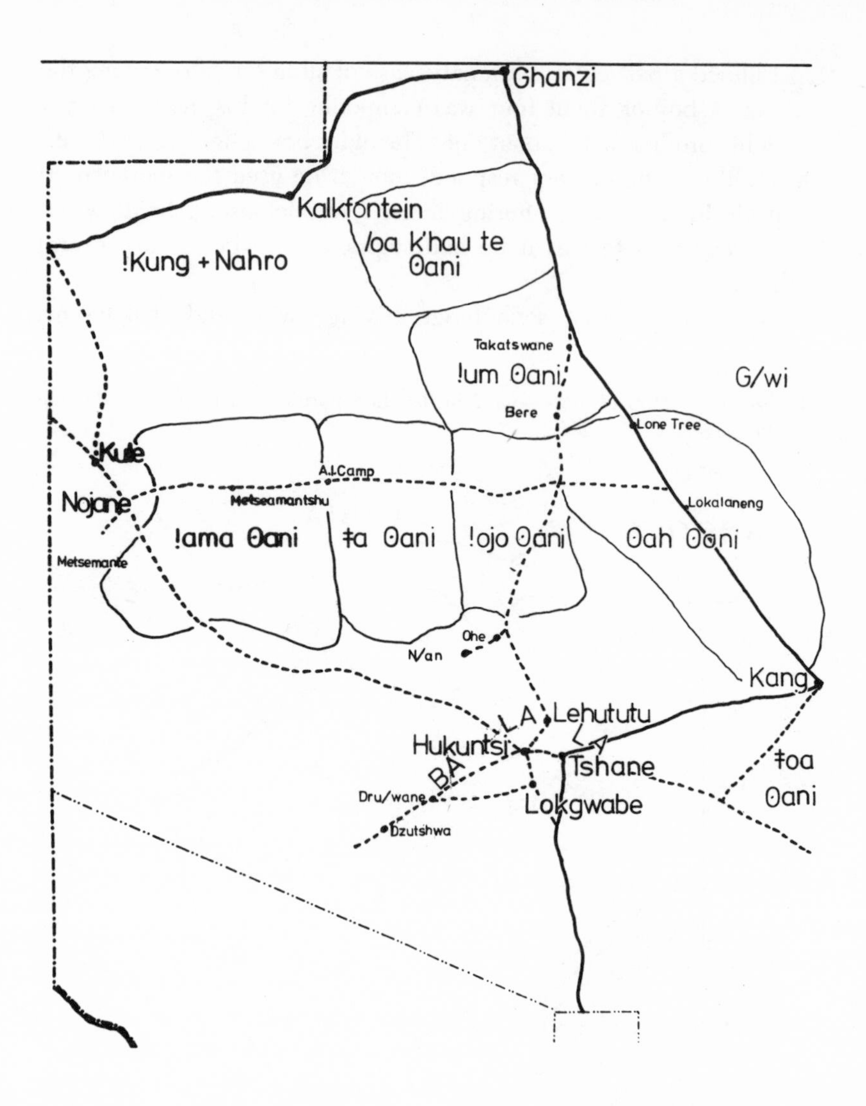

Figure 18b. The nexus territories of the !Ko Bushmen discovered by Heinz. Each nexus has a name (*miate*). The people of Takatswane are called !um Θani, "people who follow the eland antelope," or "eland people." The Okwa people call themselves /oa K'hau te Θani, or "people of the Backbone" (name for the Okwa Valley). The group in the west near Nojane are known as !ama Θani, "people among whom the sun sets." Next, going east, come the ǂa Θani, "people who (in relation to the !ama Θani) live in the east." Then come the !ojo Θani, "people who drink the !ojo pan," and finally, the Θah Θani, whose name has not been translated. The ǂΘa Θani, "the people of the south," live south of Kang. After Heinz (1966) and in press.

younger brother. He took away things with which the baby was playing, obviously hoping to anger him, for he then demonstratively threw them away. The mother was kept very busy separating them. She did this with a great deal of patience, keeping them apart with her hand and physically restraining the older boy in his attacks, but without punishing him. The smaller boy was well able to take care of himself, and here too the mother acted as peacemaker. When he picked up a stone to throw at his brother, for instance, she peremptorily held out her hand, whereupon he gave it to her. She then showed him how to play a game with the stone and handed it back to him, whereupon he began playing the game, diverted from all aggression.[78]

Fear of Strangers. At seven months, Bushman children begin to behave shyly with strangers. Fear of strangers is released both by Europeans and by strange Bushmen of both sexes. I have the impression that the reaction is stronger than it is with European children of the same age group. After the age of two, the fear of strangers gradually diminishes. My observations completely agree with Konner's among the !Kung Bushmen.[79] The strongest reaction of fear was shown by children from ten to twenty months; they hurried to their mother, clung to her, and often wept. A !Ko boy who ran away from me in fear at the age of ten months defended himself against an approaching stranger at the age of twenty months by making as if to strike him.

Aggression in Children's Play Groups. Children at play often follow behavior patterns that, since they are regularly to be observed in real clashes, might be interpreted as aggressive but for the fact that additional signals, such as the concomitant laughter, and also the course of events itself, tell us that they are "playing." Playful aggression does not lead to breaking off contact; the attacker and the attacked often change places, and strong social inhibitions prevent anyone from being deliberately hurt.

Real conflict also occurs in children's play groups. Children will slap or punch, throw sand or kick, and will often do this so violently that the victim will cry and go away. They have a strong inhibition against using objects to strike with and tend to restrict the use of sticks to threatening. The same applies to large missiles. Other aggressive actions are pushing with the shoulder with the intention of knocking the other party down, pushing with the hips, pinching, scratching, hair pulling, wrestling, and spitting. I have described these behavior patterns in detail in my monograph and provided pictorial evidence, and shall consequently not repeat them here.

Among Bushman children, there is a remarkable threat and submission behavior pattern. When they threaten each other, they clench their fists, frown, and bare their clenched teeth. At the same time, they stare at their opponent, who generally stares back. A "staring out" duel of this kind ends when one of the parties breaks off eye contact; he drops his lids, bends his head slightly, and often pouts—all of which are behavior patterns of submission. The winner then turns away; but it can often be observed that—obviously as a result of the pout—he will behave in conciliatory fashion and seek friendly contact. Many readers will recall that staring is interpreted as a threat in the European cultural area also. A hundred years ago in Germany, a man could be challenged to a duel in this way; and in other cultures, also, it is considered impolite to stare. In conversation with a single person, one looks at him but also keeps looking away; conversational partners who stare fixedly at one are disagreeable; generally they are overcompensating for an inner anxiety. There are also pathologically overanxious individuals who cannot face being looked at by their fellows and are thus inhibited in communication with them. A well-known example is the autistic children reported by Tinbergen and Tinbergen.[80] In conversation with one person, there is normally a delicately balanced interplay of looking and looking away ("optical cutoff"). Looking at a person indicates attention and readiness for contact, but it must not be overdone and degenerate into staring. Staring is sometimes used in conversation to make the other party feel insecure and to get the better of him. Underlying this behavior is an aggressive desire for domination.

It is worthy of note that the finely balanced interplay between optical contact making and optical cutoff works immediately in conversation with representatives of other cultures. I assume that a phylogenetic program causes optical contact to be interrupted automatically by the appearance of anxiety. Thus the phenomenon would have originated as an expression of successive ambivalence.

Incidentally, threatening eyes often occur in human artifacts intended to avert, and serve as a protection against, evil. Figures intended to ward off evil spirits are often characterized by a threatening stare, and protective eyes are often to be found on amulets.[81] Finally, primates and some of the lower macaques feel staring to be a threat.

The gestures of submission (dropping the head and eyelids and pouting) are also among the "universals": they are, so far as is known, to be found in all cultures.

Aggressive clashes occur rather frequently among Bushman children. In a play group consisting of seven girls and two boys, I once counted 166 aggressive acts in the space of 191 minutes. There were 96 attacks with the flat of the hand, the fist, or an object; 23 kicking incidents; 8 sand-throwing incidents, as well as other acts of the most varied kinds, such as wrestling, spitting, etc. About one-third of the acts were obviously playful. The rest led to a child's submitting and going away, and in 10 cases, the attacked child burst into tears.

Quarrels were often about the possession of objects. Boys would try to rob girls of the melons with which they were playing ball. This was frequently done as a provocation and a challenge. If the other party accepted the challenge and attacked, the challenger could respond aggressively; similarly, teasing was often used to provoke attacks, which then often served to justify to some extent the massive counterattack that followed. I never saw a child making a violent attack on another without some visible excuse. Many times attacks had an exploratory character. Smaller children in particular seemed in this way to be testing out their social field of action and their rank in relation to their older playfellows.

The oldest girl in a play group generally acted as play leader. She initiated and took charge of the group's play activity. She also made it her business to act as peacemaker and to intervene when quarreling or fighting broke out. She used her aggression educationally by punishing aggressors, and occasionally she would demonstratively carry out aggressive actions, which confirmed her in her rank. When she joined the playing children in the morning, she would regularly begin by knocking melons out of the children's hands or assuming a threatening expression and scolding them. Only then would she begin playing with them. This behavior was clearly a demonstration of the rank that gave her the authority enabling to perform her peacemaking function.

Attacks generally led to defense and retaliation. Retaliation could be planned and could take place after an interval had elapsed: I saw a boy who had been attacked go into the bush, carefully select a switch, break it off, and return to the group after ten minutes to strike his assailant with it.

Finally, a serious clash could occasionally develop out of a playful brawl, for instance, if a child involuntarily kicked another too hard.

Children often threatened each other by saying: "*Ma jonka*" ("I'll beat you up") or "*Ma kei ja*" ("I'll kill you").

Educational Aggression. Though Bushmen seldom resorted to corporal punishment when children misbehaved—more often they were scolded, when they did not share with others, for instance—I observed occasional use of corporal punishment. A boy of about twelve months followed his sister when she went to defecate and, as she did not watch him, took a mouthful. The mother saw this, hurried over and cleaned the boy's mouth with her finger. At the same time, she scolded the girl, and gave her two hard slaps on the head. The grandmother also arrived and helped in the cleaning up, and she too scolded the girl and gave her a few slaps. When a girl robbed a small boy of some meat and the boy wept in protest, his father came and took the meat from the girl and gave her a slap on the head, whereupon she threw sand at him. He returned and gave her another slap.

Finally, a notable incident that we did not see but that was reported to us: a girl of about fourteen was too familiar with various men of the group. This angered her father and elder brothers, who finally beat her.

Adult men are aggressive at initiation time. They beat the boys with switches and intimidate them. I have already pointed out the significance of this aggression (p. 88).

Children do not always tolerate adult aggression passively. I have mentioned the girl who responded to punishment by throwing sand. In another instance, a girl inadvertently upset a pot, wasting part of the contents. When her father scolded her, she lost her temper, picked up the pot, and threw it to the ground. As a result, the rest was wasted too. Her father said no more.

On another occasion, a man hid an object from an infant to prevent him from swallowing it. The infant and a boy of about six searched the man's body for it with a great deal of laughter. This developed into a playful exchange of blows between the man and the six-year-old, which gradually escalated until the man finally struck too hard, whereupon the boy ran away howling and came back with a big antelope horn and some other big bones, which he threatened to throw at the man. The boy's father, who was present, turned the situation into a joke and pacified the agitated boy, who ended by laughing again.

No attempt is made to stop very young children when they try to hit people with sticks. This is considered amusing and is laughed at, which encourages the children to continue. The immediate effect seems to be to encourage their self-awareness.

Socializing Children's Aggression. Essentially this takes place in the children's play group. Older children intervene when two younger children quarrel; they punish the aggressor and console the victim. They see to it that everyone adheres to the rules of sharing and that no one gets out of line at play. It has sometimes been argued that children should socialize themselves without adult intervention, but as Sbrzesny points out, this works only if children of different age groups are members of the same play group, so that the older children can perform educational functions.[82] This is hardly the situation in our nursery schools.

Adults intervene verbally only if an attacked child's crying goes on too long. I never saw an older child being encouraged to aggression. This was a striking fact when compared with what happens in other cultures. A Himba or Yanomami mother will not console her attacked son—or in the case of the Yanomami, her attacked daughter—but, as we have mentioned, will put a stick in the child's hand and encourage it to hit back. I saw a Himba mother vigorously striking her son when he did not respond to this admonition but went on howling. Among the Yanomami Indians, I filmed an incident in which a wrongdoer was held while his little sister was instructed in retaliatory biting and hitting. In our culture too, boys are told not to be cowards, but to defend themselves when attacked.

Bushmen stand for a different cultural ideal. They are peaceable, which does not mean that they are notably lacking in aggression. But they do not deliberately cultivate their aggressive trends. I never saw Bushman children shooting toy arrows at each other, which is a favorite boys' game among the warlike Yanomami Indians.

Aggression between Adults. Bushmen have an inflammable temperament, and though their social ideal is peaceful, quarrels ending in homicide are relatively frequent. Lee found that the murder rate among the !Kung exceeded that in the United States.[83] Heinz says that the !Ko maltreat and torment lower-ranking members of the group. Also they often threaten to kill each other. "I'll kill you with my medicine," they say. He describes the attacks of rage to which Bushmen are occasionally liable: "An angry Bushman finally settles down with a face that shows an unbelievable degree of anger. It takes very little for this anger to cause a wrestling and punching encounter with sticks and knobkerries. If the reasons are serious the fight will deteriorate to one in which knives and spears are used."[84] Jealousy often causes fights between married couples, and adultery sometimes leads to bloodshed.

Soon after our group of Bushmen moved from Takatswane to a newly dug borehole near Bere, the Okwa Valley group, who belong to the same nexus, also arrived there. The two were now living in closer proximity than usual, with the Okwa group in the territory of the Takatswane group. This led to numerous conflicts. In December 1973, there was a violent clash in which one man suffered an open head wound, another a severe bite in the hand, another a broken rib, and several others minor abrasions and bites. We had to take one man to Ghanzi for medical treatment.

Heinz reports a case of homicide: A man was playing a musical instrument that another man wanted. "You've been playing long enough. Now let me play," he said. The first man ignored the request and went on playing. When he refused a second time, the other man tried to grab the instrument and, when he failed, lost his temper and struck the first man on the head with his digging stick so hard that the stick broke; he then took the instrument and made off. Thereupon the first man took a poisoned arrow from another man's quiver, ran after the assailant, and stuck the arrow into his arm. The other members of the group pursued the assailant with bow and arrow. He escaped in the dark, but died the next day.

Bushmen try to prevent aggressive tensions from escalating into fighting. When tensions arise between two families in a band, one of them packs up and leaves the community for a time.

Efforts are made to settle strife verbally. If a man is offended, he sits outside his hut in the evening and complains. He mentions no names, but everyone in the small community knows what it is all about, and the individual so denounced is under strong social pressure to initiate a reconciliation and exchange friendly words with the offended man next day. There is also a willingness if possible to overlook offenses against good manners.

In 1972, the !Ko band that we first observed near Takatswane moved to a spring near Bere and took up cattle breeding. The transition period from food gathering to herding cattle turned out to be difficult. There were many conflicts about cattle, and quarrels between adults escalated into fighting more often than before.

It does not follow from this that property makes people aggressive. The increase in aggressivity in this case was due rather to difficulties in adaptation to a new way of making a living. Adapted herdsmen such as the Himba in the Kaokoveld of South-West Africa do not show greater intragroup aggression than the !Ko did before they started cattle breeding.

Verbal Aggression. Quarreling Bushmen abuse each other. The terms they use illustrate the tendency to dehumanization I have mentioned (p. 96), and their expletives impute physical defects, particularly in the sexual field.* Terms used by women in abusing each other are:

a maga ′i — You shit!
a sa a tshxa — Go and eat shit!
n//aba kane ka a — I don't like you!
a ba n/′a — You'll die!

References to animals are frequent in women's abuse:

a ki n/u — You're a hyena!

Insults often include references to the sexual organs:

a ǂǹ ǂgaba /i — Your penis is no good!
ke ǂa′a ǂauku bi ǂuli — Your clitoris is like a long stump!
a /anate ǂauku be chune — The lips of your vulva are as long as a baboon's!

One phrase is:

n ki dzai ma a e — I'm hungry, I'll eat you up!

Or, more freely translated:

I'm so angry with you I'll eat you up!*

* I am indebted for this information to Mrs. Elizabeth Wiley, who in 1973 became the first teacher of Bushman children at Bere.

Simpler terms of abuse, according to Heinz, are "you penis" or "you testicles" applied to a man, or "you vagina" applied to a woman. Women are also charged with having sexual relations with their father or brother.

Chaffing and Mocking. Bushmen like mocking and laughing at divergent behavior. Any victim of a minor mishap is laughed at. Once, when an old Bushman whose sight was failing stumbled over something and fell, everyone laughed. We foreign visitors were also laughed at if we behaved in a noticeably divergent way. Bushmen also laugh at each other's physical defects.

There are precise rules about who may joke with whom. Members of joking partnerships may tease each other and say things to each other that would not be permissible outside the partnership. In these joking partnerships, aggressions are acted out harmlessly.

Laughter and mockery force outsiders to adapt themselves to group norms. In the first place, behavior to which objection is taken is laughingly imitated. There are a number of specifically mocking expressive movements that are noteworthy because they also occur in other cultures. One is sticking out the tongue, a ritualized form of spitting.[85] Girls have two ways of presenting their genitals in mockery. They either directly approach the individual whom they intend to ridicule and raise their apron in front of him, or they turn away and make a deep bow, thus presenting their genitals in the primate fashion. This movement can easily be confused with the presentation of the buttocks that also occurs among Bushmen, but the nature of the bow and the concomitant behavior differs from that in the sexual presentation. Girls put sand between the cheeks of their buttocks before presenting them, and when they bow in front of the person they are intending to ridicule, the sand is released. This must be a ritualized act of defecation.[86]

Black Magic. This is used against members of a group as well as outsiders. On my last visit, I was told that a member of the Takatswane band who was angry with a Bantu had thrown bones to bring about his death (which, however, did not occur). The technique of bone throwing is the same as that used for oracular purposes.* The !Kung have also long been known to possess small

* Antelope bones are taken and thrown on the ground. The position in which they fall shows where game is to be found or what visitors are to be expected. Bones can also be thrown so as to point in a certain direction, thus conjuring up events. Sometimes the bones are spat on before being thrown.

bows that were originally thought to be homicidal weapons ("the Bushman's revolver"). Germann discovered that they were magical instruments that could be used to harm an enemy.[87] According to Dornan, !Kung Bushmen used to dance before setting out on warlike expeditions, and on these occasions, they shot their tiny arrows in the direction of the enemy or into the sun.[88] Vedder twice observed the use of a magic arrow against enemies: "In one case, a Bushman who had betrayed another was shot with an arrow in his *pontok* [Bushman hut]. When this Bushman later happened to die of a recurrence of syphilis, as was medically established, all the Bushmen were convinced that the arrow had caused his death."[89] Wilhelm reports that cattle thieves shot such an arrow against the house of a white settler, in order to cause his illness and death and thereby prevent him from pursuing them. He also saw a Bushman who suffered from rheumatism shoot himself with an arrow in both thighs. He seemed to regard the cause of his rheumatism as an enemy who could be killed by the magic power of the arrow." Bushmen also believe that people can be bewitched by poisoning their trail or their excrement.[90]

MISUNDERSTANDING AND PREJUDICE IN THE HUMAN SCIENCES

The often-repeated claim that hunters and food gatherers in general are more peaceful than people at a higher cultural level is certainly false, and so obviously false that one wonders how it can survive so stubbornly in the face of the overwhelming abundance of long-known facts. Among hunting and agricultural societies alike, there are cultures with peaceful ideals and others with warlike ideals. The Negritos of the Andaman Islands, for instance, are definitely aggressive hunters and food gatherers, and Australian aborigines used to fight a great deal, though they secured their territories in a different manner (p. 216). Schindler gives further examples of aggressive hunters and food gatherers from South America.[91]

I recently paid a visit to the Agta on the island of Luzon in the Philippines. Through our interpreter, I asked them whether they would have any objection to anyone else's gathering food on their stretch of shore, or whether their territory was marked off in any way. No, they said, they would have no objection whatever, though if So-and-So and Such-and-Such came (they mentioned some names), they would fight them, because those people had always harmed them through witchcraft, and their fathers had fought them too. The people referred to happened to be their neighbors.

Gardner has put forward an interesting theory that deserves closer examination. He believes that hunters and food gatherers who live in refuge areas tend to have peaceful ideals, while in other areas they are thoroughly warlike.[92] The thesis that primitive man is particularly peaceful as contrasted to modern man cannot be upheld, however. That this myth survives in spite of all the evidence to the contrary is in itself a matter worth investigation. One reason, I think, is the fear of the environmentalists that acknowledging inborn traits may be hazardous to their educational ideology, which assumes that man is moldable in all directions with the same ease. An example recently used to support this myth is that of the Tasaday, a cave-dwelling Stone Age people on Mindanao in the Philippines who have been described as noble, peaceful savages with no taint of aggressiveness.* The fact is that in 1971 the known Tasaday numbered 25 persons, all belonging to the same group and living together. In such a closely knit unit aggression is usually discouraged. If one were to conclude from observations of the friendly relations in a Tyrolese neighborhood that Tyrolese are a particularly peaceful people, fostering pacifistic habits, one would be wrong.

Margaret Mead is the author of the much-quoted claim that human nature is "the rawest, most undifferentiated of raw material, which must be moulded into shape by its society, which will have no form worthy of recognition unless it is shaped and formed by cultural tradition."[93] She supports this view by asserting that "the material suggests that we may say that many, if not all of the personality traits which we have called masculine or feminine are as lightly linked to sex as are the clothing, the manners, and the form of head-dress that a society at a given period assigns to either sex."[94]

Mead quotes a number of cultural models as evidence for the correctness of these claims. She attributes to the Arapesh of New Guinea a culture whose guiding educational ideal produces responsible and community-minded people of both sexes. The men are said to have little fighting spirit and war to be practically unknown. In contrast to this, the Mundugumor of New Guinea are described

* A. Montagu, *The Nature of Human Aggression* (New York, 1976); R. E. Leakey and R. Lewin, *People of the Lake: Mankind and Its Beginnings* (Garden City, 1978); J. Nance, *The Gentle Tasaday: A Stone Age People in the Philippine Rain Forest* (New York, 1975).

as violent, ambitious, crude, and immoderate, and these qualities are ascribed to both sexes, who do not exhibit more tender feelings. The reason for the difference in character is said to be early childhood experiences. The Arapesh are brought up lovingly, given a great deal of physical contact, and never subjected to corporal punishment, while the Mundugumor are harshly treated from earliest infancy. As they are unwanted by their parents, they are born into a hostile world; even the breast-feeding process is said to take place in irritation and struggle. The boys soon learn to regard other males as enemies.

Now, there is no doubt that human behavior can be culturally molded in very different ways, and that accordingly there are cultures that pursue a peaceful and egalitarian ideal while others are warlike and marked by rank. But this plasticity does not prove that behavior is the most undifferentiated of all human raw materials. Very specific and differentiated innate trends such as the aspiration for rank and aggression can be suppressed by education (whether they can be completely suppressed is another matter), but Mead's observations do not prove that human behavior can be modified with equal ease in any desired direction. Although her descriptions create the impression that this is so, the criticism is justified that her depictions overemphasize these contrasts. We know from the work of Fortune (see p. 131) that the Arapesh are not so peaceful as Mead claims; they do make war.

Without desiring in any way to detract from Mead's unquestioned services to anthropology, attention must be drawn to inaccuracies in her reports. She describes the Samoans, for instance, as happy-go-lucky, cheerful, superficial in love and family relationships, nonviolent, lacking in strict ranking order, and devoid of military virtues. In 1967, when I visited Derek Freeman at Saanapu in Samoa, he pointed out certain inaccuracies in her account:

"In Samoa the child owes no emotional allegiance to its father and mother," Mead writes.[95] But on my very first day in Samoa, Freeman showed me how the smallest boy in a family, who was weeping because he wanted to follow his mother into a boat, had to be restrained by siblings. Such scenes were repeated practically every day.

"Bravery in warfare was never a very important matter in Manua. War was a matter of village spite, or small revenge, in which only one or two individuals would be killed," she asserts;[96] elsewhere she states that warriors never had an important place in Samoan society and that courage was never highly prized. She overlooks the fact that the term *malietoa* ("brave warrior") appears on every coin and that this is the highest title a chieftain can have. It arose, according to legend, in a war with the men of Tonga, who about seven hundred years ago seized Savaii, Upolu, and Tutuila; only Manua seems to have been spared. Eventually, the invaders were driven off, and their leader, Talaifei'i, the legend says, made this appreciative farewell speech:

Malie toa, malietau	Brave warrrior, bravely fought!
'Du te le toe sau	I shall not come again
I Samoa i se aliulutau	To Samoa to make war
'A'o le a 'ou sau	But I shall come only
I aliulafalau	To make a journey[97]*

More evidence of warlike virtues is to be found in Krämer, but Mead preferred to paint a picture of an idyllic tropical island to which the problems of civilization had not penetrated. "Romantic love, as it occurs in our civilisation, inextricably bound up with ideas of monogamy, exclusiveness, jealousy and undeviating fidelity, does not occur in Samoa," she says in her celebrated *Coming of Age in Samoa*.[98]

Freeman pointed out to me that the Samoans were known as the puritans of the South Seas. They are no more steadfast in their relationships than we Europeans, but their ideals are by no means flighty, and certainly not superficial and unromantic.

There are strong prejudices in the human sciences against the assumption of genetic and phylogenetic determination of human behavior, with the result that biologists have had the most alarming denunciations hurled at them. Thus Harris classifies Darwin as a racist.[99] But since Darwin is known to have denounced slavery and to have declared the blacks in Brazil to be superior in character and physique to the white population of that country, he can hardly be charged with racism in the ordinary sense, so Harris coins the term "scientific racism." Anyone who tries to interpret race as an adaptation to an environment and examines correlations between inherited characteristics and peculiarities of behavior exposes himself to a charge of racism. Freeman has vigorously criticized this attitude.[100] But anyone who alludes to the partly genetic determination of male and female sexual roles is also denounced as a sexist.

* To pay a peaceful visit.

Those who see innate determinants in aggressive behavior are freely condemned as militarists, and those who discuss the effect of selection in the cultural field run the risk of being branded social Darwinists.[101]

But what is the real reason for rejecting the idea of biological determinants in human behavior? I think I can conclude from the literature that it is the fear that biologically determined factors are unalterable, impregnable, and uncontrollable. In fact, Luria writes of a fatalistic biologism that allegedly insists that war, criminality, and racial hatred are expressions of an unchangeable biological pressure.[102] Berkowitz, in a book that deserves to be taken more seriously on this point, expresses a similar view when he says that the Freudian theory of an innate aggressive drive "has some important implications for human conduct. An innate aggressive drive cannot be abolished by social reforms or the alleviation of frustrations. Neither complete parental permissiveness nor the fulfillment of every desire will eliminate interpersonal conflict entirely, according to this view. Its lessons for social policy are obvious: Civilization and moral order must ultimately be based on force, not on love and charity."[103]

The same fear manifests itself in the charge that ethologists seek to justify and excuse aggression, and ultimately to present it as man's inexorable fate. Rattner, for instance, writes: "In respect to politics, it is not to be overlooked that the grandiose exculpation of the problem of aggression must be gratifying to all those involved in the mass crimes of recent decades. . . . The 'aggressive drive' theory promotes a social technique of drawing a veil over things that is in complete harmony with the conservative, bourgeois way of thinking. The observer's attention is distracted from the defects of society . . . and directed to man's hypothetical 'instinctual base,' which cannot be influenced by the human will."[104]

Denker, discussing the implications of Lorenz's book, writes in similar vein: "As aggression is given a causal explanation as a natural disposition, in the view of many readers, man is to a large extent released from responsibility for himself."[105] Lumsden says: "The danger with the 'instinct of aggression' theory is, that far from emancipating man, it may enslave him to a reactionary ideology by apparently demonstrating the 'biological necessity' of an authoritarian social system organized for internal and external repression."[106]

And Lepenies and Nolte, in their otherwise very sound and readable criticism, say that recourse to man's archaic (aggressive) inheritance does not serve the cause of reflection and emancipation but that of antienlightenment. They say that those who believe man to be aggressive will set aggressive aims for him.[107] Similar criticisms appear in Selg, Hollitscher, Schmidt-Mummendey and Schmidt, and in the papers edited by Montagu.[108]

Schmidbauer garbles the issue in very crude fashion:

> It is easy to see how many ideological functions the assumption of "innate" human social behavior patterns can perform. Practically everything that persistently worries the consumer of information is explained without his being called on to make any greater efforts or to make any greater change than to detect deplorable relics of animality in man—inherited characteristics or consequences of self-domestication. War and genocide, criminality and pollution of the environment, sexual promiscuity and neurosis—all these are attributed to "instinctive" human dispositions that have gone astray. And when everyone has perceived these innate behavior trends, things will be better.[109]

In view of the repeated declarations by ethologists of the necessity of controlling aggression (see the quotation from Lorenz on p. 3), this is certainly an unfair picture.

Ethologists have never said that aggression must be accepted as man's inexorable fate. I have pointed out in a number of works that in modern conditions, phylogenetic adaptations may have lost their original function and survive merely as historical burdens, rather like the appendix. However, there is no need whatever to reconcile ourselves passively to such inheritances. As cultural creatures by nature, we are in a position to create a cultural superstructure for the control of our innate behavior patterns. Insight into their structure will help us to find the least burdensome educational strategies. Finally, those who assume that there are no inborn predispositions, and therefore pay no heed to them, are much more likely to run the risk of making excessive demands on humanity. This applies in particular to Skinner,[110] with whom I have dealt elsewhere.[111]

For all these reasons it is vital to try to understand how human behavior is constructed, and preconceived ideas must not be allowed to hamper the search for truth. In my earlier days, Lorenz always impressed on me that a scientist must be prepared to jettison one untenable hypothesis every morning at breakfast. If it should turn out that our evolutionary theory is based on false assumptions, that we are wrong to regard phylogenetic adaptations as determinants of human behavior, and that all the evidence we have accumulated from those born blind and deaf and from the comparative study of cultures and species is based on errors of perception, then we should discard our hypotheses and devote ourselves to the question of how it happens that all men so quickly and at the right time learn the correct thing to do from the point of view of preservation of the species.

It is perfectly clear to me that, in our efforts to understand the workings of the apparatus that guides our behavior, we are dependent on the very apparatus that we are investigating. It is an instrument that developed phylogenetically, and its capacities are limited. True, it reflects a real environment, but not always without distortion. Just as in the field of perception, we are subject in spite of ourselves to certain compulsions—the various optical illusions provide a good example of this—so we are subject to certain compulsions of thought. Since they are based on phylogenetic experiences, they facilitate our orientation in the world, but we must reckon with the fact that there are such things as delusions of thought, deceptions inherent in the inaccuracy of our thinking apparatus. For instance, thinking in pairs of opposites, such as *up-down, hot-cold, yes-no, either-or,* certainly facilitates thought—we mentioned the antithetical principle in the discussion of expressive behavior—but it inclines us even in scientific matters to paint in black or white, to think in terms of *either-or* rather than in terms of *both this and that.* The fact that we so often insist that the facet of reality revealed by the tools of our own discipline is the truth—the whole truth and not just an aspect of it—must surely be based on this.

The question of how we humans acquire our behavior programs can be cleared up only through interdisciplinary cooperation, since every discipline sees different facets of reality. But communication is made more difficult if those who hold differing views give accounts of ethological ideas so garbled that they amount to a misrepresentation, and then proceed to attack them. It is this that disturbs me, for instance, in the recent, otherwise stimulating contribution by Michaelis, who speaks of a "popular" or "vulgar" concept of aggression and describes the ethological position as follows:

> Forced as we are by the chronicle of devastating wars and daily reports of ever-increasing acts of violence to recognize the senselessness of aggression, it is comforting to be provided with a plausible

> and apparently irrefutable explanation of the phenomenon; namely, that a biological "drive" is responsible for the fact that we continue to commit aggression against our better knowledge. Just as nature compels us to spend a considerable part of our lives in sleep and to spend our money on food and establishing a tolerable biological climate, so it is not within our choice to live in peace with one another. The longer we go without food, the more urgently we need it, and the longer we have not committed aggression, the more inclined we are to react aggressively to the most trivial stimulus, or actually to seek out opportunities for aggressive activity.
>
> Society has reached agreements (social norms) on guiding natural compulsions into generally acceptable channels. . . . Consequently, adequate channels, which would be regarded as socially innocuous, must be established for the aggressive drive, such as sport, hunting, and "just" wars.[112]

Michaelis then attacks the idea of drives, saying that it has "long since" been scientifically demonstrated that acts of aggression are "merely a reaction to definite external stimuli," and he too attributes the alleged impact these biological ideas have on the public to the fact that they exonerate people from moral responsibility.

It is perfectly clear that a phenomenon as complex as aggressive behavior is not innate in us in the form of a complete entity and that there is consequently no point in asking whether it is innate or acquired. The question we have asked is whether phylogenetic adaptations can be shown to be codeterminant factors in the motor and receptor activity of the drive systems and learning dispositions and how these cooperate with acquired factors so as to result in a functional whole. Stating the question in this way shows that we ascribe great importance to educational influences in the formation of human aggressive behavior. Finally, we do not explain war as resulting from an innate aggressive drive. It is the result of cultural evolution, which is certainly based on phylogenetic evolution and carries it further.

In the process of cultural pseudospeciation, human groups set themselves off from each other as if they were representatives of different species. The inborn aggression controls that, in man, serve to defuse aggression, as they do in the case of animals, thus work only in intragroup conflict. Intergroup conflict assumed traits reminiscent of intraspecific conflict in animals: it became destructive. This led to a conflict of norms. The culturally imprinted norm "Kill the enemy," who, as we have said, is regarded as nonhuman, conflicts with the biological norm "Thou shalt not kill." The culturally imprinted can perfectly well suppress the constitutional, and even dispose of the relics of "bad conscience" (see p. 188). But man can also gain insight into the mechanisms involved and act in accordance with his biological propensity, and signs are discernible that development is moving in that direction. Wars are ritualized, humane conventions are developed, aggressive war is itself officially banned. Cultural evolution copies biological evolution along the path to ritualization, and this enables us to foresee the future course of development. But, as I have already briefly noted, war can develop into a contest free of bloodshed only if the defeated can escape. This stimulus situation by which fighting is switched off must be attained. But on our overpopulated planet, there are no more territories to which people can escape, and thus there can be only slight hope that the automatism of evolution will lead to a complete ritualization of war. Our only hope is for an evolutionary process guided by reason and motivated by the conscience that urges us to act in agreement with our biological norm filter.

An essential condition for development toward peace is recognition that the representatives of other cultures are fellow human beings. That does not necessarily imply an amalgamation of all cultures into a single world culture. People can be educated to an appreciation of differences, to tolerance and appreciation of different systems of values, and therefore to regarding the multiplicity of cultures as itself a value. Hence communication is necessary; we must therefore make efforts to awaken people who erect barriers to communications and demonize "enemies" to the danger of what they are doing, and at the same time we must try to immunize wide circles of the population against such indoctrination by similarly awakening them. World peace is not necessarily utopian; it is perfectly consistent with our human constitution, and we can choose it if we wish. But an additional prerequisite for it is that the living conditions of the different peoples must be internationally guaranteed—functions that have hitherto been performed by war and the threat of war.

CAUSES AND FUNCTIONS OF WAR

I am in accord with Quincy Wright's definition of war as armed conflict between groups. As I have said, war is the result of cultural evolution, and its origins reach far back into the early history of mankind. The theory of the diffusionists, who asserted that war was invented in predynastic Egypt and was copied and spread by its uncivilized neighbors, has long been discarded as obsolete. Nor does the neo-Rousseauist variation on this theme, according to which armed conflict first appeared with the development of agriculture and horticulture—Paleolithic hunters and food gatherers having been peaceful creatures in accordance with man's original nature—stand up to critical examination.

"At no period in human history was there a golden age of peace," declares Quincy Wright,[113] one of the leading experts in this field. Only if one takes a narrower view of war, regarding it, as Clausewitz does, as a rational instrument of foreign policy aimed at forcing one's will on another party, can war be regarded as a discovery of civilization. Then, as Clausewitz says, it is "the continuation of politics by other means."* Such civilized warfare assumes the existence of laws that define the states of war and peace and lay down rules of behavior for each. The difference between this kind of war and war among primitives is only one of degree, however, for among primitives too, conventions frequently exist governing the course of conflict, particularly between related tribes. At all events, war at this level is a socially recognized form of intergroup conflict that includes violence. Both the legal and the social definitions distinguish between the state of war and the state of peace.

However, there appear to be groups of primitive peoples among whom planned and organized warfare is unknown. They defend or avenge themselves merely in response to immediately preceding events. The Tasaday, recently discovered on Mindanao, in the Philippines, possess neither hunting nor war weapons. They live in small groups that seem to have little contact with one another. They avoid contact with nonmembers of their tribe. Manuel Elizalde, minister for the minorities in the Philippines, told me that a group of Tboli who went hunting in the Tasaday forests were attacked by some Tasaday. But informants had told him that they knew little about fighting and some had been killed. The reason for the attack was not known. The Tboli had set up some monkey traps in the area, and it was thought that a Tasaday might have been caught in such a trap and suffered harm. It is not known whether the Tasaday, who have scarcely been investigated, have hunting and food-gathering territories. In many respects they remind one of the Phi Thong Luang (Mrabri), who, when Bernatzik made contact with them,

* "An act of violence to force the other party to carry out our will . . . a continuation of political relations, carrying them out by other means."

also had no hunting or fighting weapons.[114] But I am unaware of any other tribe that does not at least have hunting weapons, and we know that battle-axes, weapons unsuitable for hunting, existed in the Stone Age in Europe.

War among primitives is often limited to raids in which they stalk or creep up to the enemy, using tactics reminiscent of hunting. Wilhelm, for instance, describes a raid by one group of !Kung Bushmen on another. There had been a quarrel over some game, and a man had been killed. The band that had suffered this loss sought revenge. All the men capable of bearing arms surreptitiously surrounded the other party's wharf:

> At the wharf, no one suspected the danger threatening. At sundown, the women came back singing from the fields and set about preparing the evening meal. The noise of pestles crushing food in mortars betrayed the inhabitants' presence to the enemy. Families sat down to the evening meal, talking and laughing. The last glow of sunset faded, and the darkness of night hid everything. Here and there a camp fire flared up again. A baby cried in its sleep.
>
> Meanwhile, the enemy, creeping as noiselessly as snakes, advanced from all sides, though the onslaught was not to take place until first light. The ring closed tighter and tighter around the wharf. Here and there an inhabitant was still awake, and the dogs had not yet gone to sleep and must not be disturbed. The night passed quietly, and the zodiacal light heralded another day. By now, everyone was asleep, and even the dogs forgot their alertness and gathered for comfort around the glimmering embers. The enemy crept nearer and nearer, and at last dawn began to break. With a yell, they fell with raised assegais on the unlucky sleepers. Here a man startled from his sleep made a grab for his weapon, but was quickly struck down. There a man with the courage of despair defended himself with his assegai, but the enemy's superiority in numbers was too great, and he soon fell, pierced by many assegais. Women frantically seized their children and tried to flee, but were slaughtered without compunction. Here a mother nearly managed to escape with her baby, but an arrow pierced her side; in her agony, she threw the child away and tried to struggle on, but her strength failed, she collapsed groaning, and her pursuers approached with a bestial howl. A few blows with a kiri smashed the child's skull and finished off the mother too. Only a few lucky ones managed to get away and reach a friendly wharf. The sun rose red in the east and its first rays lit up the horrid scene. The victors were in possession of the wharf and had started looting. Everything useful was taken away. Clay pots were smashed and the huts set on fire. Heavily laden with their booty, the Bushmen set off for home. A dog, not understanding what could have happened to its master, went on

> howling in the distance, and then the silence of the dead prevailed. Soon the first pallid harriers were hovering around the scene of the massacre; vultures followed, and at night, hyenas and jackals gorged themselves on the bodies of the slain. If any of the few survivors returned a few days later, they would find that all that was left of their kith and kin was some scattered bones. But for them, too, the hour of vengeance, when like would be paid for with like, would one day strike. Thus warfare continually prevails between hostile clans and tribes. The misdeeds of the individual have to be paid for by all.[115]

Warmaking by the Aranda tribes of Central Australia was described by Strehlow:[116] The first step was for chieftains who wanted to fight to send messengers to friendly chieftains inviting them to join in an alliance. With them, the messengers took a rope made of dead men's hair, a bone worn in the nose, an eagle's feather, and a small churinga.* The message was immediately understood, and if the group did not wish to attract the hostility of its friendly neighbors, there would be a men's assembly, at which the decision to help them would be made. The chieftain would then return the objects to the messenger, who would go on to the next group. Other warriors invited to join in would gather where the chieftain who was first invited lived. Before they left to join the chieftain who issued the invitation, messengers would be sent to him to announce that they were coming. The chieftains and their men followed, and fires would be lighted to announce their approach. The chieftain who was expecting them would then order his warriors to adorn themselves. They painted their breasts and belly with black stripes edged in white, and the approaching warriors decorated themselves in the same way. The warriors of different groups greeted each other by calling out "Wa wa wa bau." Then the host chieftain would say: "Let us sleep here for one more night. In the morning, we shall set out to strike the enemy." During the night, the warriors would put themselves in a pugnacious frame of mind by singing war songs. Strehlow translated some of these:

> Where have they come from who are going a long way?
> They have painted themselves very black.
>
> Strike his penis with the pointed bow.
> Put the bone at the edge of the penis.

* A sacred board on which the symbols of the tribe's totemic ancestors and their history are engraved in highly stylized form.

Blood flows like water from the long penis
Over the shoulder of the man sitting in front.

Into my own spear thrower,
I put the spear, I put the spear.

The spear thrower throws the spear, the spear thrower throws,
The spear strikes the foe, the spear strikes.

The barbed spear
Mangles the enemy.

The spear whizzed, it whizzed,
The spear crashed.

It struck him fatally, it struck him fatally,
He cannot remove the spear from his wound.

He crashed to the ground like a *jack-arro,**
He collapsed like the sky.

We extract their entrails and eat their fat
After we have removed the skin.
We tear their entrails.

The songs refer to a ritual that the warriors carry out on the night before the attack. They open their circumcision wound† with sharp bones and let blood flow over each others' right shoulders. This is done to make the right arm strong. They also engage in a kind of self-indoctrination to familiarize themselves with the idea and the situation of killing. The song also includes cannibalistic threats.

The next morning, the men gather and continue the process of putting themselves in the proper frame of mind. They play with a spear thrower and boast of being invulnerable. They talk themselves into believing this; it is obviously a way of overcoming their fear. Then they march off, fully armed. They spend a night near the enemy's encampment and attack immediately before daybreak when everyone is asleep. Before the attack, the chieftain gives every warrior a piece of rope made of dead men's hair and places a bandicoot tail in the mouth and right armband of each of them in order to put fire into their bellies and make them better able to strike down the enemy. Each warrior paints a white stripe on his brow and the sides of his nose. Then they creep up to the enemy's camping place and surround it. First, crying "Wai, wai, wai," they spear the

* Butcherbird (*Cracticus*), which swoops like an arrow on its prey.
† The Central Australian tribes cut along the length of the urethra (subincision).

sleeping men. Then, crying "Kukukukuku," they kill the women with cudgels, and finally, they deal with the young children, grasping them by the feet and smashing their heads on stones or on the ground. After completing this murderous work, they slit open the bellies of the slain and eat a little of the raw stomach fat. They make a circle around the slain and leave them, unburied. At a water hole they wash the blood from their spears, and the young warriors drink the mixture of blood and water to make themselves strong.

The Maoris also gave no quarter to defeated enemies. A hundred years ago, an old Maori reported to a Western observer: "When once the enemy broke and commenced to run, the combatants being so close together, a fast runner would knock a dozen on the head in a short time; and the great aim of these fast-running warriors . . . was to chase straight on and never stop, only striking one blow at one man so as to cripple him, so that those behind should be sure to overtake and finish him. It was not uncommon for one man, strong and swift of foot, when the enemy were fairly routed, to stab with a light spear ten or a dozen men in such a way as to ensure their being overtaken and killed."[117] Dead enemies were eaten.

War among the Maoris was a form of vengeance for robbery and other crimes. Men, however, seem to have learned at an early stage that massive retaliation can boomerang. At all events, conventions for combat developed early—at the cultural level of hunters and food gatherers and at the level of Neolithic horticulturalists. Special rules reduced losses on both sides. Contests took place in which sticks were used, for instance, in the case of the Hadza. Fighting among Australian aborigines often took place on these lines. Lumholtz vividly describes an intratribal clash in north-east Australia. The opposing parties confronted each other adorned as if for a feast and armed with spears, clubs, boomerangs, and hard wooden swords; their wives and children came with them. Then the attacking party charged, to the accompaniment of savage war cries:

> The strange tribes on the other side stood in a group in front of their huts, which were picturesquely situated near the edge of the forest, at the foot of the scrub-clad hill. As soon as our men had halted, three men from the hostile ranks came forward in a threatening manner with shields in their left hands and swords held perpendicularly in their right. Their heads were covered with the elegant yellow and white topknots of the white cockatoos. . . . The three men approached ours very rapidly, running forward with long elastic

leaps. Now and then they jumped high in the air like cats, and fell down behind their shields, so well concealed that we saw but little of them above the high grass. This manœuvre was repeated until they came within about twenty yards from our men; then they halted. . . .

Now the duels were to begin; three men came forward from our side and accepted the challenge, the rest remaining quiet for the present.

The common position for challenging is as follows: the shield is held in the left hand, and the sword perpendicularly in the right. But, owing to the weight of the sword, it must be used almost like a blacksmith's sledge-hammer in order to hit the shield of the opponent with full force; the combatant is therefore obliged to let the weapon rest in front on the ground a few moments before the duel begins, when he swings it back and past his head against his opponent. When one of them has made his blow, it is his opponent's turn, and thus they exchange blows until one of them gets tired and gives up, or his shield is cloven, in which case he is regarded as unfit for the fight.

While the first three pairs were fighting, others began to exchange blows. There was no regularity in the fight. The duel usually began with spears, then they came nearer to each other and took to their swords. Sometimes the matter was decided at a distance, boomerangs, nolla-nollas [clubs], and spears being thrown against the shields. The natives are exceedingly skilful in parrying, so that they are seldom wounded by the first two kinds of weapons. On the other hand, the spears easily penetrate the shields, and sometimes injure the bearer, who is then regarded as disqualified and must declare himself beaten. There were always some combatants in the field, frequently seven or eight pairs at a time; but the duellists were continually changing.

The women gather up the weapons, and when a warrior has to engage in several duels, his wives continually supply him with weapons. . . .

The old women also take part in the fray. They stand behind the combatants with the same kind of sticks as those used for digging up roots. . . . They cry to the men, egging and urging them on, four or five frequently surrounding one man, and acting as if perfectly mad. The men become more and more excited, perspiration pours from them, and they exert themselves to the utmost.

If one of the men is conquered, the old women gather around him and protect him with their sticks, parrying the sword blows of his opponent, constantly shouting, "Do not kill him, do not kill him!" . . .

With the greatest attention I watched the interesting duels, which lasted only about three-quarters of an hour, but which entertained me more than any performance I ever witnessed. . . . Boomerangs and nolla-nollas whizzed about our ears, without however hindering me from watching with interest the passion of these wild children of nature. . . .

After such a conflict the reader possibly expects a description of fallen warriors swimming in blood; but relatives and friends take care that none of the combatants are injured. Mortal wounds are extremely rare. [One man] had received a slight wound in the arm above the elbow from a boomerang, and was therefore pitied by everybody. In the next borboby [duel] one person happened to be pierced by a spear, which, being barbed, could not be removed. His tribe carried him about with them for three days before he died.

As soon as the sun had set the conflict ceased. The people separated, each one going to his own camp, all deeply interested in the events of the day. There was not much sleep that night, and conversation was lively round the small camp fires. As a result of the borboby several family revolutions had already taken place, men had lost their wives and women had acquired new husbands. In the cool morning of the next day the duels were continued for an hour; then the crowds scattered, each tribe returning to its own "land."[118]

Rules of combat understandably developed first between closely connected ethnic groups, among whom it was relatively easy to reach an understanding. The Kiwai of Papua, for instance, make a clear distinction between related village communities and those of different tribal origin, who are regarded as hereditary enemies. They always try to kill the latter, but in intratribal fighting, there is little bloodshed. To European observers, these battles, which are fought with a great deal of noise and fury, look exceedingly dangerous, but fatal wounds are rare. The tribesmen bombard each other with sticks and arrows, but they aim at the legs, and headhunting is not practiced.[119]

The battles of the Dugum Dani in the Baliem Valley in West Irian, the Indonesian part of New Guinea, are also conducted according to rules that prevent excessive bloodshed. Certainly the dead are counted up, and for everyone killed on one side, somebody must be killed on the other. The aim is to achieve parity, and when this has been reached, fighting ceases. But since the other side generally believes itself to be in arrears, feuds drag on for years, and though the number of deaths per incident is low, they mount up over the years.[120]

Alliances are formed between groups of villages. Each group inhabits an area of about thirty-five square miles, and the whole of the Baliem Valley, with its roughly 50,000 inhabitants, is divided up among several dozen allied groups. Each is marked off from the rest by a frontier and a stretch of no-man's-land. Tall watchtowers are built on pales along the frontiers, and on top of each, a man keeps watch all day long. The alarm is given if enemies are seen approaching. The watchtowers are placed at intervals that a runner can easily cover in five minutes.

The wars of the Dugum Dani are not fought for territorial gain. The chief motive they give is the necessity of avenging the spirits of those who have been killed. If they failed to do so, they would be plagued with illnesses and other disasters; floods or drought would afflict their crops.

Men are prepared for their role as warriors from an early age; they are experts who engage in war almost as a form of sport. In fact, they can do so with a good deal of confidence, because a fatal casualty in their battles tends to be something of an accident.

Their battles are of two types: the formal pitched battle and the raid. A challenge is issued for a formal battle. A group of men go to the frontier in the morning and call across to the enemy. The challenge is nearly always accepted, and the news is passed on. The men of neighboring villages may decide to fight on the same day. They put on all their finery: some hang ivory tusks on their noses, others decorate their heads with bird-of-paradise or white heron feathers, and they all smear their bodies with pig fat. The warriors gradually assemble on the traditional battleground; there is complete agreement that the battle can begin only when both sides have completed their preparations. Gardner and Heider describe the event:

> By noon, most of the warriors have arrived and the various formations have taken more or less final positions. Some are armed with bows and arrows, some with spears. The opposing armies are deployed so that between the most forward elements of each there lies a battleground of perhaps five hundred yards. A mood of silent but excited expectation pervades all ranks. From this point on, the day will bring the pleasures of the fight to several hundred on both sides, momentary terror for the handful who will feel the sudden pain of an enemy arrow, and, rarely, the unmentionable shock of death to someone who acts stupidly or clumsily.[121]

The warriors wait with increasing tension for a party from one side or the other to advance onto the battleground and thus open the battle. The other side responds by advancing onto the battleground in cautious stages. When the two sides are about fifty yards apart, a few arrows are released and the parties withdraw again.

After several sallies of this kind, which represent a kind of preliminary skirmishing, the battle begins; both sides advance onto the battleground, rhythmically stamping their feet, and for about ten or fifteen minutes, there is an engagement involving one to two hundred warriors. Arrows and spears fly through the air, but they are adroitly avoided. Sometimes, when there is sufficient cover to enable them to make a flanking movement unobserved, one group succeeds in attacking the other from two sides at once, but generally such moves are detected and the attempt fails.

In their efforts to avoid the arrows and spears, and provide cover to comrades who have thrown their spears, the warriors are perpetually on the move, and their place is continually taken by others, giving them an opportunity to rest. Gardner and Heider were struck by the fact that, though bird feathers play a big part in the Dani culture, their arrows are unfeathered, which makes their flight very erratic. "A possible explanation might be that the Dani realize that more warriors would be hit if their arrows were feathered. Perhaps they realize that even such a minor alteration in the rules of combat would disturb the fine balance they have established between chance and performance," they speculate.[122]

In a single day's fighting, there may be from ten to twenty hits. Hostilities cease in the evening, enabling everyone to get home before dark. Before the two sides separate, they spend a long time mocking and insulting each other. As they all know their opponents by name and also know all about their private lives, there are a great many personal references, which causes a great deal of laughter on both sides.

In addition to this kind of warfare, the Dani also practice raids, which Gardner and Heider contrast to the pitched battle: "A raid is a desperate attempt to take an enemy life and has none of the theatricality of a formal battle."[123] Specially experienced warriors are selected for raids. They try to creep undetected into hostile territory and surprise someone there. But they are sometimes detected and ambushed themselves.

Raids take place when the group has failed to secure compensation for a loss by other means. Each side reckons up its dead, which leads to the perpetuation of feuds. Gardner and Heider estimate that the annual losses amount to about ten to twenty deaths on each side. That does not represent a serious threat to the survival of any of the groups. Functionally, these demonstrations of combativeness on the frontier seem to be a ritualized form of territorial demarcation.

Many other tribes have ritualized their clashes as the Dani have. Bonds of kinship generally diminish aggression. Layard described the levels of escalation between different groups of the inhabitants of the New Hebrides. Fighting between members of the same clan is carried out only with wooden clubs, and serious efforts are made to reconcile the parties. In clashes between villages on the same side of the island, clubs and spears are used, but the encounters are very formal and take place on battlegrounds specially set aside for that purpose. In wars with inhabitants of neighboring islands, the object is to take the enemy by surprise, but the *lex talionis* applies: the dead are counted up and the object is to secure parity, which prevents excessive bloodshed. But this law does not hold good in clashes between totally alien groups.[124]

Armed conflict is frequently associated with cannibalism. The motives vary. The dead are often regarded as food. One of the Jalé of West Irian assured Koch that they would eat a dead enemy "because he tastes good, as good as pork, if not actually better." No special inhibitions are felt about the dead, unless they are known personally. The Jalé say: "People whose face one knows must not be eaten."[125] Profane cannibalism used to be rather widespread; we have examples from South America, Africa, New Guinea, Australia, and Oceania and several other areas.[126] In certain environmental conditions, cannibalism actually plays a part in group diet, since many groups suffer from a great shortage of high-grade protein.[127]

Another motive for cannibalism is revenge. In this way, the victor achieves total destruction of the enemy. In the Solomon Islands, the greatest humiliation that could be inflicted on an enemy was to eat him. Cannibalism out of hatred is known to have existed all over the world.

Finally, cannibalism is sometimes motivated by magical ideas; men believe that by eating an enemy, they can acquire some of his qualities—his courage or strength, for instance.*

Men's attitude to cannibalism varies. The extensiveness of profane cannibalism in past times suggests that a dead enemy was regarded simply as a piece of meat. The dead seemed to lack important signals

* Magical ideas also underlie the various forms of endocannibalism, in which dead relatives are eaten. I shall not go into this subject here.

that release impulses such as sympathy. Only at a higher level of reflection does man become aware that his dead enemy was human too, and inhibitions against cannibalism then arise.

If we desire to eliminate war, we must first establish whether it performs functions in interhuman relations, and if so, what functions. I have already stated my view that war is the result of cultural evolution, whose direction is determined by selection. In accordance with this, war should demonstrably contribute to the preservation of a culture. This conflicts with the view that war appears only as a phenomenon concomitant to other functional systems or as a morbid degeneration of them. Walsh puts forward the highly speculative view that fathers send their sons to war to punish them for their oedipal wishes.[128] According to this theory, war would be the result of paternal hatred without having any other function. Fromm distinguishes between beneficial and malignant aggression. The former is described as a phylogenetically programmed impulse to attack whenever vital interests of the individual or the species are threatened. This defensive aggression contributes to the preservation of the individual or the species, and it ceases with the disappearance of the threat. Malignant aggression is characterized by destructiveness and cruelty and is based on sadism and necrophilia—character traits that occasionally appear as morbid developments in man.[129] *

Certainly, sadism exists as a pathological phenomenon, and it is often responsible for the atrocities ordered by many rulers. Apart from these, it offers no explanation whatsoever of the phenomenon of war, in which atrocities do sometimes occur, though they do not play an essential role. In many wars, chivalrous rules are actually observed.

Discussion of the phenomenon of war suffers from a certain confusion in the use of the terms "cause" and "function." In speaking of the cause of war, we sometimes mean the immediate occasion for it, or the reasons for it advanced by some people. The term is also used in an evolutionary sense, referring to the selective pressure that "caused" a development. It is this that I mean when I look for the "function" of a development, and I shall be guided in my search by the observable consequences of warlike conflicts.

Another point on which misunderstanding occurs in the contro-

* Fromm, like many others before him, attacks the theory allegedly put forward by ethologists that war is caused by an innate aggressive drive, an inborn tendency to kill. Ethologists, however, have never suggested this.

versy about function became clear to me in reading Hallpike. He criticizes the functionalists among anthropologists who assume that war among primitives performs a function, and who take the view that an institution exists only because it performs a task and hence is also necessary. He talks in this connection of a "functionalist illusion" and gives some examples to show that institutions are not always adaptive. I do not know whom he expects to surprise by this; he certainly does not surprise biologists, who have long pointed out that loss and change of function can appear in the evolutionary process, and that, in changed environmental conditions, structures that once fulfilled a function are carried on merely as vestigial survivals. That certainly applies to cultural institutions also. Demonstrating that in some societies an institution such as war performs no—or no definite—function does not mean that it never performed a function. Hallpike summarizes his conclusions in the following four points:

1. The fact that an institution exists does not mean that no other would have sufficed in its place.

2. The fact of its existence does not mean that its existence was necessary.

3. The fact of its existence does not mean that it was the best possible institution in the circumstances.

4. The fact that an institution is necessary for a society to survive does not mean that it will necessarily be formed.[130]

Points 1 and 3 say practically the same thing. Biologists have considered this question. Kramer asked whether nature has always produced the best possible technical solutions, and concluded that it has not always done so. On the contrary, it has made "constructional errors" simply because in the course of evolution, use was made of whatever happened to be available. Air-breathing terrestrial vertebrates developed out of fishlike ancestors whose blood circulation had to be reconstructed in adaptation to terrestrial life by means of the secondary insertion of a pulmonary circulatory system. The reconstruction took place in several steps; consequently, in amphibians, the circulation of venous and arterial blood is only incompletely separated. The mixture diminishes efficiency. In a completely new development, better technical solutions would have been possible.[131] I can agree completely with points 1 and 3, but points 2 and 4 have no heuristic value, since they are propositions that cannot be tested. Whether a structure that actually came into existence might also never have done so seems a question hardly worth quarreling about, and whether an institution necessary for survival will really develop is not a question worth pondering.

Finally, it is hard to understand Hallpike's surprise at our inquiring into the function "even of such an institution as war." War is, after all, a very noticeable, very widespread, and very ancient phenomenon that can hardly be expected to be neutral from the point of view of selection. The possibility of its being detrimental or useful to the preservation of culture must surely exist. If the former were always and invariably the case, counterselection would long since have set in. But that, as history shows, is by no means the case.

Let us, therefore, take a look at the consequences of war. For the losers, these have in very many cases undoubtedly been catastrophic; one only has to consider what happened to the Tasmanians or the North American Indians. The winner makes territorial acquisitions, of which there are innumerable examples right down to the present day. New settlement areas and mineral resources become accessible to him; he can spread and multiply and make up for his war losses. Selectively, this has certainly tended to encourage breeding for aggressivity. Fromm maintains that this cannot be so, since the more aggressive simply destroy themselves by exposing themselves more.[132] This is an argument that is heard occasionally, but it betrays inadequate information about the mechanism of evolution. As we have mentioned, this takes place in closed groups, in which the genome of the killed survives provided the group survives, for the members of the group are their close relatives; and in the long run, survival is the reward of groups that produce the bravest men, who are most ready to fight. The defeated are exterminated, driven away, or subjugated and thus culturally annihilated. In the latter event, their idioplasm is generally absorbed by the victors. For a long period of human history at least, war favored the selection of fighting spirit and aggression. But, as Bigelow pointed out, in addition to fostering the military virtues, it also encouraged intelligence and the ability to cooperate in intergroup competition.[133]

The history of mankind down to the present day is the history of the successful conqueror. Whether or not territorial gain plays a part in the subjective motivation of war is a completely secondary question in that respect. What counts is the result.[134]

I emphasize this because Chagnon, in his otherwise admirable studies, repeatedly insists that the warlike Yanomami Indians do not aim at territorial gain: "Territorial gains are neither intended nor achieved in the carrying out of these conflicts. This has certain consequences for the theories of aggression based on territorial behavior, particularly in the form developed in the recent books of Ardrey and Lorenz."[135]

Although the declared objectives of their wars are to capture women and to show other groups that they are ready to defend their sovereignty by force, nevertheless, the demonstrable result, apart from the capture of women and the gain in prestige, is that the winners often exterminate the losers or force them to abandon territory. It is this result that counts, even though the motivations put forward by those involved are different. In other cultures, men go to war to distinguish themselves, but the result is the same. The selective advantage for the group is independent of the individual's frame of mind. A young man courting a girl does not by any means invariably connect this with the intention of becoming a father of a healthy baby as quickly as possible. Wright states this clearly: "The function of an activity may be broader than its intention."[136]

We have seen that many tribes in New Guinea fight in accordance with fixed ritual. According to Rappaport, wars among the Tsembaga of New Guinea develop out of individual quarrels. If anyone marries a woman without securing her relatives' consent, or commits rape, or if a pig belonging to another group wreaks havoc in a man's garden and he kills it, or if he steals field crops, or is suspected of working black magic on a member of another group, efforts are made to kill the evildoer. This in turn calls for retaliation, and warlike conflicts often ensue. The more highly populated an area is, the more probable such conflicts are. Rappaport points out that if twenty men each have a pig and a garden, there are 400 chances of a pig's causing conflict by damaging a garden, but if forty men each have a garden and a pig, the chances of conflict increase to 1,600. The same applies to other possible causes of dispute. In other words, the chances of conflict increase faster than the increase in population; if the population increases in arithmetic progression, the chances of conflict increase roughly in geometric progression.[137]

According to this, increasing irritability and tension would lead to conflict long before overpopulation developed. First of all, the enemy's weaknesses are probed in long-lasting feuds; here we see a phenomenon comparable in function to intragroup exploratory aggression. If the irritability and tension increase, serious clashes will occur, which may end with one group's being driven away. But an alternative possibility is that things may settle down to a state of equilibrium—if two groups, through birth control, for instance, keep their population at the same level. In such cases, rules can develop that make coexistence possible, with war tending increasingly to become a frontier demarcation ritual in which no blood is shed.

The territorial function of war has been plainly seen by many anthropologists.[138] According to Morey and Marwitt the centralized tribes of the South American lowlands engaged in wars that served to acquire land and raise tribute and provided an answer before the days of European influence to the problem of population pressure on a limited area of cultivable land.[139] The aggressivity of the groups had an obviously ecological basis.

The endemic warfare of the Yuma tribes in the Colorado and Gila river areas in the United States can also be interpreted as a struggle for the cultivable land. It was only the flat country along the rivers that was fought for; groups that lived in neighboring territory and had a different style of life were not regarded as competitors. Thus the Mojave Yuma did not make war on the Yarapai, the Western Apaches, the Chemehuevi, or the Cohuila who lived on the neighboring hills.[140]

The small closed groups of Maoris, in New Zealand, reacted with war to the slightest real or imaginary provocation by their neighbors. But clashes were generally brief, and though the victors tried to kill as many of the vanquished as possible, the total was not excessively high. The most important consequences of war were that groups split up and the pressure of their neighbors enforced the cultivation of new land. This adaptive system collapsed when Europeans introduced muskets, whose use led to heavy losses. Although the losses had certainly been considerable before then since the victors gave no quarter and generally slaughtered the fleeing enemy, after the introduction of muskets, they became catastrophic.[141]

Layard reports a similar state of affairs from the New Hebrides, where the introduction of muskets led to wholesale massacres, causing escalation in even the normally formalized conflicts between neighboring villages on the small islands off the coast of Malekula.

> In this way, during the latter part of the nineteenth century the small islanders practically wiped out the whole population of what was once a flourishing district, containing innumerable villages immediately inland from the adjacent Malekulan coast. The sites of

> these villages, including the mainland village of Tolamp, are now pointed out in what is thick jungle. Not only were the mainland villages thus decimated, however. The same tragedy occurred even in warfare between two of each individual small islands. The first villages to acquire muskets were in all cases those situated on the "superior" side of each island, which, being in possession of the best beaches, were first to come in contact with white men. These muskets they then used against the members of the other side of their own island, with the result that the villages on the "inferior" side were severely handled. This was the case with the two villages of Emil Marur and Emil Lepon Atchin, both of which were nearly wiped out. The more far-seeing of the small islanders now bitterly regret these suicidal ravages, which have so seriously reduced their numbers in the face of the growing menace of the whites [Layard was in the New Hebrides in 1917 and 1918]. But the measure of their regret still depends on the degree of kinship ties between themselves and their victims. For, while muskets are now with common consent banned in warfare against their fellow small islanders, they are still used against the few remaining inhabitants of the adjacent mainland.[142]

Wars are fought for hunting grounds, pasture land, and arable land, and if in earlier times, climatic alterations made a group's living area inhospitable, it was actually compelled to find new territory by force of arms. The drying up of the Central Asian steppes set the Mongol peoples in motion, and their warlike expeditions took them all the way to Europe. Their clash with the Teutonic peoples in turn forced the latter to migrate. Overpopulation, too—as the consequence of technical or medical discoveries, for instance—can force human groups to migrate.

The fact that wars are about territory has often been clearly recognized by those involved: "But of the cities of these people, which the Lord thy God doth give thee for an inheritance, thou shalt save alive nothing that breatheth: But thou shalt utterly destroy them; namely, the Hittites, and the Amorites, the Canaanites, and the Perizzites, the Hivites, and the Jebusites; as the Lord thy God hath commanded thee" (Deut. 20: 16–17).

"And they utterly destroyed all that was in the city, both man and woman, young and old, and ox, and sheep, and ass, with the edge of the sword" (Josh. 6: 21).

It was perfectly clear to the lawgiver who proclaimed these things that his people needed their neighbors' land as a settlement area. Since men normally have strong inhibitions against aggression directed at women and children, this massacre dictated by cold utilitarian considerations had to be represented as a divine command. Such commands, however, have not always been obeyed. With the further development of civilization, humanitarian considerations came increasingly to prevail. Victors satisfied themselves with reducing their enemies to subjection, extracting tribute from them, and imposing their own culture on them. Also, their labor power, for which there had been no use at lower cultural levels, came to be appreciated. The conquered were then wiped out not biologically but culturally. In this way, cultures gradually developed their own dynamism independently of their agents. And cultural formulas were able to demonstrate their usefulness and impose themselves independently of their biological agents.

War is a means that aids groups to compete for the wealth essential to life (land, mineral resources, etc.). It has also been said that it serves the purpose of keeping population growth in check, but that is certainly a secondary effect. It has been claimed that it serves to regulate psychological variables (the abreaction of psychological tensions): here individual motivations are confused with selective advantage.

If one asks whether modern war still performs functions I have described, the answer is yes, up to the present day, insofar as it has been conducted with the methods in use up to the Second World War. Modern warfare also leads to the acquisition of land and access to raw materials and labor. There are divergent views about the outcome of a nuclear war—whether it would be worth the victor's while—but the most vivid imagination is hardly sufficient to visualize the resulting devastation.

War is to be attributed neither to degenerate, misdirected animal instincts nor to necrophilia nor to any other pathological degeneration of basic human impulses. It is not a functionless deviation, but a specifically human form of intergroup aggression that helps human groups to acquire land and natural resources. We gladly ignore these unpleasant facts. We want to live in peace, and in accordance with the principle of "What ought not be, cannot be," we delude ourselves and shut our eyes to the problem. That only makes the inevitable awakening more unpleasant. It is surely better to see plainly that war performs definite functions and face up to that truth. The fact that it performs these functions does not mean that they can be performed in no other way. Better solutions can be devised, but they presuppose that the functions of war are apprehended in nonwarlike fashion. No one can expect that nations reduced to poverty because they are refused access to vital raw materials will tolerate that situation passively. No one can expect that people living in a country growing more and more inhospitable because of a change of climate will wait for death by starvation without doing anything about it. In such situations, a group will sooner or later take positive action; its only alternative is to reconcile itself to its fate and go under. If the nations of the earth want peace, they must plan further ahead than the magic year 2000 and ensure that not all empty spaces are populated and overpopulated, so that they can be allotted to those who in the foreseeable future will be compelled to migrate by climatic changes. Also a world organization will have to see to the fair distribution of raw materials. That is a matter to which I shall return. It is certain that today we are still far from reaching rational solutions. The use of force is still worth the victor's while, as was shown not long ago in Cyprus. But that is no moral justification for it.

7
On the Way to Peace

WAR AND CONSCIENCE

> When we called on Major Cuellar at his house this morning and were asked to breakfast with his family, we said to his wife: "Do you know what your husband has to do?" "Yes," she replied tersely. "Do you know that your husband can kill hundreds of thousands or actually millions of people with his rockets?" "Yes," she said. "Does the idea disturb you?" "No, not much, it's his job, after all." Then we said to Major Cuellar: "Doesn't the idea oppress you that you're really always waiting here only to fire your rockets and to kill many people and destroy another country?" "No," Major Cuellar replied, and Captain Gillespie confirmed this with a nod. "I know that if I had to turn the key here, it would be after my country had been attacked. I should only be defending my country. So I shall destroy what I have to destroy" (Hugo Portisch, *Friede durch Angst* [Peace through Fear]).[1]

War performs functions. Nations expand by war, acquire territory, reduce others to subjection in order to exploit them, and enforce access to sources of raw materials. They spread their culture, or important aspects of it, such as religion. Frontiers are protected by warlike means, and fighting consolidates the group. But though war is beneficial to the winners, we cannot accept it. Is it fear or conscience that makes us search for peace? Both fear and conscience play a part. Fear of nuclear self-destruction certainly prevents the world powers from going to war with each other at the present time. This "peace through fear," as Portisch calls it, may suffice for a time, but it offers no security in the long run. Nevertheless, it is a powerful incentive to search for solutions. These might, however, consist of such effective protection against nuclear attack and the consequences of radioactive pollution that war would again hold out the prospect of gain and thus become worthwhile. Fear is, consequently, an insecure foundation for peace. But is conscience a better foundation? The commandments innate in us "Thou shalt not kill" and "Thou shalt not steal," which are actually obeyed by some primates (see p. 70), can be effectively overcome by superimposed cultural norm filters.

We are capable of persuading ourselves that other people are not human and of acting accordingly (p. 122). If this does not work, we can rid ourselves of responsibility by claiming we are acting under orders, and turn what we are doing into a job. Or aggression can be justified by calling it defense or retaliation, in which case it can be exalted as a glorious and heroic thing almost without conflict. Few scruples are felt about defending one's group. A threat is obviously so powerful a releasing stimulus that it overrides all scruples.

Hence, nowadays we no longer indulge in wars of conquest, but only defend ourselves—or we liberate our fellow men and thus perform a good deed. In addition to the "Our country is threatened" cliché, we also misuse other clichés, such as justice and liberty, equality, fraternity, and we actually fight for peace. Land grabbing as the professed motive for aggression would not arouse the right kind of enthusiasm, so Hitler represented the German invasion of Poland in 1939 as a necessary act of self-defense and laid emphasis on the vain efforts he had made for peace:

> You know the endless attempts I made for a peaceful clarification and understanding of the problem of Austria, and later of the problem of the Sudetenland, Bohemia and Moravia. It was all in vain. . . .
>
> In my talks with Polish statesmen . . . I have formulated at last the German proposals and . . . there is nothing more modest or loyal than these proposals. I should like to say this to the world. I alone was in the position to make such proposals, for I know very well that in doing so I brought myself into opposition to millions of Germans. These proposals have been refused. . . .

> For two whole days I sat with my Government and waited to see whether it was convenient for the Polish Government to send a plenipotentiary or not. . . . But I am wrongly judged if my love of peace and my patience are mistaken for weakness or even cowardice. . . . I can no longer find any willingness on the part of the Polish Government to conduct serious negotiations with us. . . . I have therefore resolved to speak to Poland in the same language that Poland for months past has used toward us. . . .
>
> This night for the first time Polish regular soldiers fired on our own territory. Since 5:45 a.m. we have been returning the fire, and from now on bombs will be met with bombs (speech, September 1, 1939).

When defense of the group is involved, killing is regarded as a noble deed. Bravery and willingness to risk one's life for the group count heavily and win in the conflict of love and hate. Nevertheless, the biological norm filter is not eliminated by this process, and consequently, we experience a conflict of norms. All these justifications with whose aid men excuse destructive aggression do not prevent the stirrings of conscience, particularly after the aggressive emotion has faded. This may seem implausible in view of the horrors that come to our awareness daily and fill the history books. Even women and children are sometimes killed, though this generally ranks as an appalling deviation that the chroniclers describe as an atrocity. The inhibitions against killing women and children are certainly stronger than those against killing conspecifics of the same sex.

Atrocities committed under the influence of emotion must be distinguished from those that take place by order of authorities. The latter can be based on cold-blooded considerations and be ordered for the deliberate purpose of intimidating a subject population. The Assyrian king Ashurnasirpal II (883–859 B.C.) is known for the brutal frankness with which he describes the atrocities he inflicted on his prisoners. His monuments show how he blinded captive kings and mutilated them in the most varied ways. He had his victims flayed and hung on the city walls, he impaled seven hundred prisoners outside the city gate and others on the towers of the walls, and he does not fail to emphasize that they were still alive. He walled up twenty prisoners inside his palace, beheaded others, and made a pile of their skulls.[3] There have certainly been conscienceless rulers who got drunk on power and sadistically enjoyed their acts of cruelty, but cold-blooded considerations of expediency generally underlie acts of terror. This has been shown by

Edwards in analyzing the bloodthirsty terror with which Oliver Cromwell intimidated the Irish during the Puritan revolution. To the present day, the Irish use the expression "The curse of Cromwell be upon you," but the figures show that the campaign in which Cromwell conquered Ireland was the least bloodthirsty in its long history. As Edwards explains:

> The deliberate massacre of forty-two hundred men, two-thirds of them English, was his solution of the problem. By that action he subdued the island in less than nine months. He lost only a few hundred of his own troops, and three large Irish armies, then in the field, dissolved from mere terror as soon as the Puritan army approached them. If Cromwell had conducted his campaign according to the usual methods, he would have had to fight all three of the Irish armies. There cannot be any doubt that he would have been victorious, and there is equally little doubt that he would have killed at least thirty thousand Irishmen and lost probably ten thousand of his own troops, while the war would have lasted two or three years. Cromwell claims in his official reports that his policy resulted in a great saving of life.[4]

Edwards adds that Cromwell saw to it that enough Irish escaped the massacre to tell the story of it all over the country. It is noteworthy that Cromwell thought it necessary to justify his action, which points to a conflict of conscience. There are other examples of this in history. Helena Valero, who spent many years as a prisoner of the Yanomami Indians, reported qualms of conscience among these warlike people in connection with a massacre. She describes how warriors reproached themselves after a raid in which they killed some women. "Shamatari-tushaua was not a bad man. On the way he said: 'Why did you kill all these people? You should not have killed so many.' The men answered: 'You yourself said we should kill them all.' 'That was only a manner of speaking. There were only few men there.' But the others said: 'There are only few now. A large number were out hunting. They still have wives, from whom they will get more children, and then they will be numerous again.' "[5] She goes on to say that those who had killed lived apart for some time, were given special food, and were allowed to talk to no one. Later, among the Namoeteri, another Yanomami tribe, she saw warriors who bathed daily and rubbed themselves with rough, sharp leaves "to purge themselves more quickly of their misdeeds." I have already mentioned that such atonement rituals have frequently been described; Freud explains them as the expression of a bad conscience.

In a paper published in 1870, I found the story of a man who had taken part in a massacre of Bushmen:*

> "I often shudder," a worthy field cornet said, "when I think of one of the first scenes of that kind at which I was present in my youth. It happened when I was beginning my militia duty. I was in a detachment under Karel Kotz. We had attacked a big kraal of Bushmen and mowed them down. When the fire ceased, we found five women still alive. After long consideration, it was decided to spare their lives, because one Boer needed a woman slave for this, another for that. The unlucky creatures were ordered to tramp on ahead of us, but it soon turned out that they hindered our march, because they could not move quickly enough. Orders were given to shoot them down. The scene that followed still often haunts me. As soon as they realized our intention, the helpless victims sprang at us and clung so hard to some members of the party that for a long time it was impossible to shoot them without endangering the lives of the men to whom they were clinging. Four were eventually disposed of, but the fifth clung so desperately to our comrade that it was impossible to tear her away, so eventually we gave in to his plea that we should take her with us."[6]

Examples of this kind could be multiplied. It is known that soldiers drafted into firing squads have drugged themselves with alcohol and that many of them later needed psychiatric treatment. Their consciences pricked them, though they could claim to have had no choice in what they did, since they acted under orders. Men can act brutally in certain circumstances, but their conscience

* A continual state of strife existed at the time between the Boer newcomers and Bushmen long established in the area. The newcomers, who were having a hard struggle for life as cattle breeders, considered the Bushmen a nuisance who threatened their very existence, and with good reason, from their point of view, since the Bushmen regarded cattle feeding on their land as their property and killed and ate them without inhibitions. Attempts to reach a peaceful settlement had had little success. Lichtenstein writes: "Four years previously, to satisfy them completely, a herd of more than 1,600 sheep and 30 cattle had been collected as a voluntary gift from the inhabitants of the whole of the northern district to enable them to manage in an orderly fashion, pasture their herds, rear young, and begin a regular way of life. But the attempt was entirely fruitless. For, as they live without a government, without fixed dwelling places, without a social contract, and even without personal property, their distant fellow countrymen soon arrived and helped them to eat everything up until nothing whatever remained." (*Reisen im südlichen Afrika* [1811; reprint ed. 1967], p. 183).

pursues them. Certainly, there are also men to whom love, sympathy, and conscience are alien, but they are mentally ill.

When a cultural norm filter is superimposed on a biological norm filter and thus comes into conflict with it, we are aware of it, and the ensuing disharmony causes us discomfort. As the innate is stubborn and puts up greater resistance to modification than the acquired, we are subjected to pressure to bring the cultural and the biological norm filters into harmony. We can obey a cultural commandment when it conflicts with a biological commandment, but conscience reminds us of the conflict. Conscience will always conflict with a cultural command to kill, and if, as we hope, world peace is ever established, it will be based, not only on rational utilitarian grounds, but also, and much more powerfully, on our innate norms.

The Christian norm of love of one's enemy can be regarded as an attempt to adapt the cultural to the biological norm. That we give preference to the biological norm "Thou shalt not kill" is immediately obvious. As it is innate, it accords both with our feeling and our reason, so we do not question it. There is also a rational case for arguing that settling disputes by destructive wars is harmful in the long run to the human species. Moreover, increasing facilities for communication make the cultural norm "Kill thine enemy" harder and harder to maintain. Numerous contacts, even though they take place by way of such technical media as television, make us aware that foreigners are human too. Over and above our cultural divisions, we humans have remained the same, right down to the details of our repertoire of innate behavior patterns. Ethologically we are a single species and thus have a common frame of reference within which we can meet and understand one another. It is only culturally that we have been able to describe others as nonhuman, and that is something we can no longer do with a clear conscience. Through many contacts with people of the most varied cultures, a binding sense of common humanity has developed.

But what happens if our intelligence speaks for a cultural norm while our feelings—the subjective counterpart of an innate norm—say the opposite? How are we to decide between conflicting biological and cultural norms? Is there a ranking order of norms?

The phenomenon of conflicting norms occurs in the biological field. The motivations that cause an animal to care for its young can come into conflict with others. If, for instance, an enemy seems too powerful to resist, self-preservation will take over, and the animal will flee instead of defending its young. Also, hunger can sometimes be so predominant that a mother animal will neglect its young or even eat them. In these cases, the animal is put into a state of conflict between the norms by which its behavior should be governed, between flight and attack. Which impulse then prevails in the parliament of instincts depends both on the animal's mood and on the strength of the releasing stimuli.

In man, as I have observed, obedience and sympathy often come into conflict, and the experiments by Milgram mentioned above show that sometimes obedience prevails and sometimes sympathy, depending on the proximity and status of the obedience-demanding authority. Similarly, conflict sometimes occurs between sympathy with outsiders and intolerance of them. Unreflectingly, we give preference now to one, now to the other. There is no more a strict ranking order of norms than there is a strict hierarchy of drives. In both cases, the ranking order is approximate. All the same, understanding in this field enables us to establish a ranking order and, for instance, place tolerance higher than intolerance.

To clarify this, let us briefly consider values and the establishing of norms in the cultural field. There is a very extensive philosophical literature about the relationship between innate norms and culturally transmitted ethical norms; the attempt to explain the binding nature of norms is very ancient. Now anthropologists have described a multiplicity of different cultures and systems of value, and hence at the present day often tend to a cultural relativism, regarding the norms by which men direct their lives as relative and specific to their culture. This is indeed true of many cultural norms. But biological norms are universal.

The attempts to work out a system of natural law, like the attempts of the Catholic church to read God's will from nature—a subject on which Wickler has written with great clarity[7]—show that in general we regard the universal norm as imposing the stronger obligations.

The question arises whether it might not be possible to develop universally binding cultural norms too. Kant tried to lay the foundations of a rational morality by asking the test question: Could the principles that guide this individual's conduct be elevated into a general law? Stated in more general terms, the question is whether a particular item of behavior is or is not disturbing to social coexistence. It can also be reformulated into questioning the value of conduct from the point of view of preservation of the species: Would it or would it not be detrimental to the human species if everyone acted as I am about to act now? A narrow definition of the human species, restricting the privilege of complete "humanity" to one's own group, leads to the development of norms directed only to the survival of the group and devoid of universal validity. But if one defines the human species biologically, including all the races that compose it, the norms developed will have universal validity. The cultural norm of the preservation of the species will have a higher place in the scale of values than norms that are culturally relative. The latter can be allowed to stand so long as they do not conflict with the interests of humanity. In fact, for reasons that I shall set out shortly, the right to live one's own cultural life within the framework of humanity should be better assured than it has been in the past.

An evolutionary process guided by reason based on the question of survival value would be in harmony with the principles that govern biological evolution through the mechanism of mutation and selection. That part of cultural evolution that is not guided by reason would also be tested by selection for its contribution to the preservation of the species. The process involves risk, however. Developments that fail to prove themselves, that do not stand up to selection, are "errors," which many cultures have paid for by going under. But rationally guided evolution can preserve us from such errors.

In our pluralist society, therefore, our intolerance of outsiders is by no means adaptive. Society has received many impulses from brilliant outsiders. Our biological inhibitions against aggression, aided by this insight, should enable us effectively to control our reaction to outsiders. In the conflict between the two biological norms that we experience as intolerance and sympathy, understanding helps us give greater weight to the latter.

Survival value is the binding guiding principle, but for the time being, it remains an open question whether one should have in mind only the survival of a definite culture or racial group or of the whole of humanity. In the last resort, this is an ethical choice. The fact previously emphasized that, over and above all cultural differences, we are alike in every detail of our innate behavior shows that biologically we are still a species, and that consequently the preservation of humanity must be our aim. A rational argument is that nuclear war would imperil the whole of humanity. The argument that there would be no victors is well founded.

With the aid of the question of the preservation of the species, the further evolution of humanity could be guided by reason and, to a large extent, freed from the unintelligent system of learning by trial and error. This too would involve risks, of course, since the system that has prevailed hitherto of blindly probing all possibilities is a tried and tested mechanism of the evolution that creates the "hopeful monsters" by whom new paths of development are opened up. Humanity must preserve a certain adaptive variability and keep trying the new. From that point of view, the complete amalgamation of cultures that is often advocated is questionable, since the elimination of multiplicity would restrict evolutionary prospects. It seems reasonable to bear this in mind and to tolerate culturally different systems of value so long as they do not offend against the overriding principle of the preservation of the human species. A humanity that unites humanity must not be absolutely uniform, but must be united with tolerance, so that new ideas may be tried out and have a chance of proving themselves. Otherwise the evolution of man could come to a standstill.

A MOURNING RITUAL IN THE HIGHLANDS OF NEW GUINEA

In August 1972, there was a fight between two Mbowamb-speaking tribes, in the course of which three men were killed. The Jika had taken possession of some land belonging to the Yamaka by cultivating it at night. This had happened several months previously, and an amicable settlement of the disputed border issue was to have been brought about by an exchange of territory. The Yamaka duly surrendered land to the Jika, who did not keep to the agreement, but mocked the Yamaka.* On September 11, they wounded a Yamaka, and with that the conflict escalated. Next day, four hundred Yamaka and six hundred Jika met in battle, and one Jika and two Yamaka were killed before police put a stop to the fighting. By the afternoon, all the men had been disarmed and most had been arrested. On September 16 and 17, I filmed the Yamaka mourning ceremony at Tega,† a sing-sing place about five miles east of Mount

* All this is the Yamaka version of the story.
† Dieter Heunemann and my son Bernolf helped with the filming.

Hagen. Mourning had been in progress for two days, but it was only on the day of our arrival that the police handed over one of the bodies. In the course of the mourning ceremony, the whole episode was dealt with and deplored in speeches and song. Since these speeches provide an insight into the mentality of a Papuan tribe and its attitude toward war and peace, I shall reproduce them and describe the ceremony.

First, a few words about the Mount Hagen tribes. The Leahy brothers came to these highlands in the early thirties to search for gold. They made the first contact with these tribes[8] and constructed an airfield near what is today known as Mount Hagen, aided by a representative of the Australian administration. They were followed in 1934 by Father W. Ross, a Catholic missionary, who still lives there, and the Lutherans established missions there too. Pacification of the area was interrupted by the outbreak of the Second World War and was not resumed until after 1945. The inhabitants have now to a large extent been converted to Christianity, but they have not been forced to give up their old practices. Many of them work for the government and for everyday purposes have adopted European clothing, manners, and money.

For their *moka* rituals (gift exchanges), the *tanim hed* (a courtship ritual),[9] and mourning rituals, they wear the traditional costume that is still worn by country people. It is perhaps a characteristic of progressive tribes that they do not throw their own culture overboard completely; instead, a healthy self-esteem causes them to preserve their heritage without cutting themselves off from progress. This positive evaluation should, however, not blind us to the negative consequences of European influence—especially alcohol.

When we arrived at Tega on the morning of September 16, a mourning group had already gathered in the middle of the sing-sing place. The son of one of the dead men was crouching on the floor with lowered head, lamenting aloud. Several women were crouching around him, grieving with him and comforting him, stroking his body and hair. Around this group were men and women who had smeared their faces and the upper parts of their bodies with yellow clay. They wore skirts made of green cordyline leaves. Some women wore Western-style clothing, and there were a few men in shorts. The women carried cordyline branches, and the men had long, pointed sticks instead of the spears that had been confiscated by the police. They sang songs in chorus in praise of the dead man, saying how handsome he was and how well he had tended his fields.

Other guests kept arriving in small groups. The circle of mourners broke up and men and women formed into two separate groups. After putting themselves into the right frame of mind by vigorously stamping their feet and calling out "Uaeh uaeh," the men, led by two men dancing in line in front of them, advanced on the guests, brandishing their sticks. They were followed by the women. The behavior of the men suggested a simulated attack. The women who followed them waved their green cordyline branches. In this way, threat-greeting and peace-greeting were combined.[10] The warriors and the women circled around the guests and danced back to the mourning group, some of them guiding the guests. The latter lamented loudly when they reached the mourning group. Men and women took hold of their hair as if to tear it out, and a number of men did the same with their beards. Some of the women wept bitterly, their tears leaving distinct traces on their clay-smeared cheeks. Everyone expressed his sympathy in this way. Finally they embraced and comforted the dead man's family. The young man whose father had been killed was in the middle. Whenever new mourners arrived, he rose and lamented aloud. This went on for two days. In the early afternoon, groups of men formed, and one of them spoke while the others, crouched on the ground, made comments and applauded. Each speaker walked excitedly up and down, and when one had finished, another took his place.

The second day followed the pattern of the first. In the afternoon, guests arrived with presents for the mourners. At about five o'clock, the coffin with the dead man's body was brought out of the hut. The mourners sat on the ground in front of it while a prayer leader read a Christian prayer. The coffin was then taken away for burial.

A minor incident that day was finally resolved satisfactorily. A man who resented our presence removed an attachment from my camera case and told me I could have it back as soon as I took my departure; he said that our filming disturbed him. Through our interpreter, we explained that we were making a scientific film for universities and schools, and that we were not there simply as inquisitive tourists to satisfy a craving for sensationalism, but to learn about the customs of other people. He then gave me back the part and was very friendly. He was followed by a number of men and women who wanted to dictate their views directly onto our tape. These too were highly informative.

TRANSLATIONS OF SPEECHES RECORDED UNOBSERVED DURING MOURNING CEREMONY TAPE NO. 7*

FIRST SPEAKER: Jika tribe is no good and so is Yamaka tribe [*tape inaudible because of noise made by mourners*]. The other tribes have killed man but I did not expect this tribe [Jika] to kill him. Now I am in the sun crying for the victim, but I will not take revenge. It's all right with me that my man got killed and I expect pay-back ceremony instead of fight.

SECOND SPEAKER: The Jikas and Yamakas are fighting and the whole place and people wandering around like they don't own houses or pigs, etc. And now most men have gone to prison and that makes things even more worse, so, brothers, stop this and let's settle properly. The white man has brought us law and order, so let's live in peace. I am sorry for the people, the women and the children, so please, brothers, let's stop the fight and let's live in peace.

THIRD SPEAKER: The fight has taken place.† I heard this when I was a boy and now I see here in Hagen the fight between Jika and Yamaka tribes and this is going to ruin the law and order brought in by the white man [government]. My people, bordered by the Gumach ringi,‡ and me are coming from far away. We hear this bad story about a fight right in the town where the administration center is, and how can you town people expect us keep law and order? Now the fight in the heart of Hagen broke out and all men gone to jail. I wish to ask the other tribes around Hagen, Moges, Kimis, Kulis, Kuklikas, Kelis, etc., to take good care of women and children. I who am speaking am from Dei council area and I am just pouring out my feelings, so it's all up to the neighboring tribes to do something about it. I can only say things and not do anything because I come from far away. So all us outside tribes are getting an idea that all tribes in and near Hagen are fighting, and not the Jikas and Yamakas only. The fighting tribesmen have left all belongings, women, and children and gone to prison, so I want all men who are near and have not been involved in the fight to take good care of them.

FOURTH SPEAKER: I appreciate what you have said [*addressing third speaker*]. I will try and do my best. You are from Dei council, and I,

* The speeches were translated by Martin Wimb, a Mbowamb, under the supervision of W. Straatmans of the Australian National University at Port Moresby. The tapes are filed under these numbers at the Max Planck Society Human Ethology Film Library.

† The reference is to past times.—I. E.-E.

‡ I was unable to discover the significance of this evidently geographical expression.—I. E.-E.

from Hagen council here, will try to solve it. I also got the same feeling as you . . . [*tape unintelligible because of mourning by new arrivals*].

[*The mourning song or cry is as follows:*]

Sun rising from the east. And I hear, I hear the story of victim. If this man is my man, if so, who shall I live with? Oh my man, who shall I live with? O my man. [*This song or cry goes on to the end of the tape.*]

TAPE NO. 8

FIRST SPEAKER: The Yamakas are easy tribe. They have been troublemakers, but now among the big tribes, it is only a small tribe. I see that the Jika tribe is a bit of show-off, because the Jika tribe is as big as the Moges and should [compete with] the Moges. Why did it fight the Yamaka, which is only a small tribe? I see that the Jika tribe is afraid of the Moge tribe and turned around and fought the small tribe. Now listen to me, the men have left their wives and children, pigs, coffee trees, and everything, so every tribe that hasn't taken part in the fight must take care of the things left behind by the men who went to prison.

SECOND SPEAKER: You [from the] Menembi, you [from] the famous and well-known tribe giving the speech. I believe you. What you have said is all true. For fighting other things you are *the man* and [of the right] tribe but I believe law and order is something real. The government is here and we are going to be self-governing soon. Don't worry about things left behind by the men who went to jail, because they'll be safe. See now the Jika and Yamaka fought yesterday, but now [they are] in prison because law and government are very powerful things. So what we should talk about now is how to settle [the case] properly and keep law and order. . . . [*Tape unintelligible because of several simultaneous speakers until third speaker begins*]

THIRD SPEAKER: You people listen to me. The man Makura is lying dead here. You other people are just pouring out false speeches. The dead man belonged to my tribe, and it's not a small one. So my brother Makura is lying dead here. And you young men giving speeches are not saying what I wanted, so I am the old man giving speech. [*A man in the background saying:* Police are coming so please settle down! *Suddenly the mourners begin weeping and calling out as the police approach. This continues to the end of the tape.*]

TRANSLATION OF SPEECHES RECORDED AT SPEAKERS' REQUEST
TAPE NO. 9, FIRST HALF

FIRST SPEAKER: Walimulk Nori [leader of the fighting Jika tribe] has claimed the Montila land already. Now he wants to get Punthulk land. I have given him land already and more. I got a bit cross when he wants Punthulk land because you [*going over to direct address*] have claimed all my land and where do you want me to go? That's how the fight started. I went for a fight over Puntumbulk land and Walimulk Nori has killed Makura and I am now sorry again. People who know the victim can imagine what sort of man Makura is but [for] those who don't know Makura: [he] is a man whom the tribesmen respect and honor but now he has been killed and I am telling the story that is being taped.

SECOND SPEAKER: Now I am going to make a law. This is what I did when the councillor was here. You might ask what I am going to do? I do such things when the Europeans or white man or natives want to take me to court or something, I do this. And people feel OK. I am going to give food. When the councillor was here, we used to give food to such people and now I am going to do it without the presence of the councillor. So now I want any expatriates or natives to listen and take in what I say. What I do the Jika tribe doesn't do and no other tribes do. When the big officers come at four o'clock, as Moge Paia has arranged, I will give food, so Europeans and natives from outside should come and see. That's all.

[*A cry song. There is a man introducing it:*]

All men have gone to prison and I left myself only behind because I was sick. I am going to put a cry into the radio [meaning the tape]: Dad, I don't see that man who killed you. This man has killed and gone away among men. Oh! Dad, dad, whom shall I live with?

A SPEAKER: My councillor who assists me in doing such things is gone to prison, so I am alone and I am going to give chickens [as a mourning gift], which I bought to the policemen. I spent $30. If the councillor was here, we should have done a better and bigger one, but I am alone now and this present is not as big as it would have been in the presence of the councillor. That's all, my name is Jaki.

TAPE NO. 9, SECOND HALF

From our village, all Yamaka boys and men have all gone to jail. There are no men in our village now; only all us married women of the Yamaka clan are gathered here. No men to do the men's job, instead of men crying and greeting the relatives who come to take part in the mourning. What men usually do in such ceremonies is now being done by us married women. We are sorry for those men who went to prison. We are now going to sing our mourning cry.

My name is Vunt and my father's name is Muramul. Here is the cry:

Dad, who wanted to own Punthulk land,
eh–eh–eh–eh–
Your child is sad and lonesome, eh– What shall I do?
eh–eh–eh–eh–
Dad's footprint is going west to Punthulk
eh–eh–eh–eh–
The enemy men were good with their hands
eh–eh–eh–eh–
Dad who wanted to own Punthulk land,
eh–eh–eh–eh–
Your child is sad and lonesome, eh– What shall I do?

[*Repeated again and again.*]

That's enough. If I were a man, I would have fight with the tribesmen and go jail just the same, but I borned to be a woman so I am now suffering and suffering for my father and crying.

Now two wives of my brothers are going to cry:

In our village, there are a lot of men who obey law and order. Us women who are married to these clansmen and sisters of the clans are always happy at all times. Now this trouble arose but our councillor did not have a bad reputation. And with his unexpected trouble the councillor has gone to Madang for imprisonment and us women who are married to his [the councillor's] sons are suffering a lot for him and are going to cry. I am the wife of one of the councillor's sons. My name is Kagle from Ronidan.

[*Kagle speaks alone:*]

Jika, Nori, Nema, and Kalimba [all leaders of the enemy tribe] have claimed and taken all Yamaka land. We [Yamakas] have been given pigs he has taken. And they should be satisfied but now they have turned against us and killed Pangal. Why they were hungry and killed him, we don't know. We never neglected them. Now men who were at home have gone to jail.

Previously, the Jikas and the rest of Yamakas fought and went to prison but no Yamaka men were involved. Now Mappulk Pangal got killed and all the clansmen have gone to jail, every one of them.

What men usually do in such ceremonies is now being done by us women, and I believe the visitors will go back and tell people of the strange performance by us women. Us women greeting and mourning for the deaths. The *kiap* [government] is here so we will not fight back over the deaths again. We are still thinking about

Kul Ponya, who is going away to Tembeka; who has killed our man and is going away to Tembeka. That is all.

[*Another mourning cry:*]

My name is Rangil. Enough.
Hello man Kainip, I should have put him in a special container.*
eh–
Hello, my husband, eh–husband, husband, who shall I live with?
Man of Walimil, Noki killed and gone to Madang eh–eh
I have only heard the story. Oh–eh–eh.

[*Repeated again and again.*]

MAN SPEAKING: Me as a Yamaka–a big-name Yamaka–work and obey law and order. I have been baptized by Lutheran mission and all people come to live together in villages composed of lined-up houses.

Me as Yamaka have welcomed arrival of the white man. I have been glad of the peace that the white man brought. I have not settled properly but since the white man arrived, I am living well and appreciate the better ways of living but Jika Nori is spoiling me [meaning his whole Yamaka tribe]. Trouble has just started between Jika Nori and me Yamaka Kera. Trouble has erupted from where Yamaka tribes have originated. I was seeing peace and multiplied the population of Yamakas with plenty of sons. Now Nori's tribe killed my first son and I am the father mourning for the first son. The victim Pangal has been killed because of Jika Nori and Nema's fault. Now the trouble or fight has erupted and my brothers and sons have gone to jail and I am suffering for them. Me Yamaka Nang saying this. Enough.

ANOTHER SPEAKER: My name is Yamaka Kundump Kumbati. My brother has been killed so I am going to mourn:

Son, eh– Kul Porya has killed son eh–, son eh–, son eh– Rakim Bkong has killed. Son eh–, son eh–, son eh– If the first son were here, he would avenge it. Son eh–, son eh–, son eh– Son, they have killed without any revenge in return. Son eh–, son eh–, son eh–

[*The whole group joins in and repeats again.*]

Both the speeches that we recorded unobserved and those recorded with the speakers' knowledge clearly reflect a wish for peace and order. Europeans appear in this context as bringers of peace. When the Australian government pacified these warlike tribes after the Second World War, it was done without the use of force; per-

* I.e., I should have taken better care of him.–I. E.-E.

suasion and firmness on the part of the government mission and the government patrols were sufficient. It is worthy of note that relatively little hate is expressed in the mourning speeches. The conflict is deplored, a settlement is offered on payment of compensation, and there are few traces of a desire for blood vengeance.

Moreover, it is clear from the speeches that visitors of different tribal origin not directly concerned deplored the conflict and reproached those involved in it. What they are saying is: If you, who live so close to the government, live like this, how can distant tribes be expected to keep the peace? There are continual references to the suffering caused to women and children. One speaker calls on friendly tribes to take good care of the women and children while their husbands are in prison, but it is pointed out to him that this is unnecessary, as the government sees to law and order. One can read between the lines that the Yamaka are too proud to accept such aid. Some slight mistrust may also be involved.

What I consider remarkable about the statements is that they show plainly that the clash was about the possession of land. It is also worth observing that the recordings reveal a way of feeling and thinking very similar to our own in the high value that is attached to peace.

CONTROL OF INTERGROUP AGGRESSION

Ritualization of Warfare

There are signs that human intergroup aggression may be beginning to move in the direction of eliminating bloodshed.[*] The process offers a remarkable parallel to the ritualization of aggression in the animal kingdom: rules are developed that prevent the parties to a war from suffering excessive losses. They prescribe how war is begun—through a formal declaration—so that the enemy is not caught unprepared. Agreement is often reached on distinguishing between combatants and noncombatants. The unarmed are not attacked, and fighting between armed men can be made less lethal by refraining from the use of every available means of destruction and using only those means whose destructiveness remains within bounds. The reader will recall that the Dani use unfeathered instead of feathered arrows, thus reducing the number of hits. Other rules specify how one can surrender and how prisoners are to be treated. Peacemaking, reparations, and reconciliation are also governed by conventions, and man finally develops other patterns of solving conflicts, so that war increasingly becomes a last resort.

The so-called civilized nations are not the first to develop alternate modes of dealing with disputes. Among the Murngin in Australia, ritualized battles take place whose object is to settle conflicts without bloodshed and thus provide a firm foundation for peace. If a group has injured or killed a member of another group, and enough time has passed for feelings to have subsided, the group seeking satisfaction sends a messenger who announces that it is now ready for a *makarata.* If this is agreed to, an encounter duly takes place. The warriors paint themselves white and draw up opposite each other, holding their spears. As a security measure (in case the battle should escalate), both sides try to pick a site with bushes to cover their rear. The challenging group dances toward its opponents, singing songs referring to its totemic ancestors. This group then turns and goes back to its starting positions. The other side does exactly the same. When both have returned to their starting lines, the battle begins. Friends who encouraged the killer (without being directly involved in the killing) then advance onto the battleground; each is accompanied by two men who have close relatives in the other group. Fear of killing a friend prevents spears from being thrown too violently. These men also try to strike down the spears to prevent them from hitting those at whom they are aimed. Moreover, the spears are blunted by the removal of their stone tips. The aggrieved members of the challenging clan throw their spears one after the other; those who feel sufficiently aggrieved do so many times. While the spears are being thrown, other tribesmen shout insults at their opponents, who must not reply. Finally, the old men of the challenging clan decide that enough is enough.

Next, those directly involved in the killing advance into the arena, and spears still equipped with their stone tips are thrown at them. But the old men tell the throwers to be careful to avoid taking life. Those on the other side also ensure that feelings do not get out of hand, telling their men to accept the insults quietly and not to use their spears, since they are in the wrong. When the aggrieved group finally has had enough, both groups dance toward each other. The killer receives a stab in the thigh with a spear, and that is the end of the matter. As soon as he has recovered, he can enter hostile territory without fear. If the wound is slight, however, it indicates that the other group still wants revenge. As soon as the wounding has been carried out, the group dances as a single body, showing that peace has been restored.[11]

Rappaport[12] has described highly ritualized warfare among the Tsembaga in New Guinea. Warfare between two groups of Tsembaga generally results from a quarrel between individuals during which one of them has been killed or injured. Such an incident may lead to war, or there may be a peaceful settlement.

If the two men belong to different subclans of a clan occupying a common territory, a bloodless resolution of the conflict is probable. When the closely related clans of the Tomegai and Merkai were drawn up in battle array, members of a third clan interposed themselves between their shields, told them it was unseemly for brothers to fight, and urged them to break off the battle, which they did.

I have already observed that the intervention of third parties as mediators and peacemakers plays a big part in individualized intragroup conflict. In the resolution of intergroup conflict, men obviously hark back to behavior patterns that have proved themselves in intragroup conflict and perhaps correspond to an inborn pattern of reaction. Fraternal warfare between groups of Tsembaga is prevented when a close relationship between groups has been built up by numerous marriage ties. An interesting consequence of exogamy is that it establishes ties to other groups that help to prevent conflicts. Various authors have regarded this alliance-creating function as having been responsible for the development of exogamy. Among the advanced cultures, alliances have certainly been reinforced by marriages between ruling families.

Peaceful settlement of a conflict among the Tsembaga is doubtful, however, if it involves neighboring groups that do not have a right to uncultivated land belonging to a common territory—even if friendly relations normally exist between the two groups and they are connected by marriage ties. Because they do not share land, there are fewer of these ties, and consequently fewer channels by which mediation can take place. Also the two units are more distinctly separated from each other, so that a clash between them does not represent a serious threat to the internal order of both. When there are many ties of kinship to the other group, a conflict would involve a substantial disturbance of its internal harmony, as relatives would find themselves on opposite sides.

Wars between groups are conducted on a basis of reciprocity. Every death on one side has to be paid for by a death on the other, and peace is improbable until parity has been attained. Pauses in the hostilities occur, but they are in the nature of a truce. Since parity is hard to achieve, feuds are protracted, and as Rappaport puts it, each round in the battle lays the basis for the next. Nevertheless, there are devices that help to circumvent this law and thus help to damp down conflict. When an enemy is eventually killed, each of a number of men can claim the killing for himself and thus regard his obligation as having been wiped out. Also, magical practices can take the place of killing. On the other hand, hostilities can escalate if responsibility for a killing is attributed to a whole group, thus involving persons previously uninvolved.

Fighting groups consist of members of local groups among whom the conflict broke out. They are joined by allies recruited from other local populations. These are generally related by marriage to members of the warring parties. If one of these allies is killed, responsibility is not attributed to the enemy, but rests with the group with whom he was fighting, which has to supply their ally's dependants with a wife whose first child bears the dead man's name. The fact that the enemy is not held responsible for an ally's death certainly helps to prevent escalation of the conflict.

Wars generally begin with a "skirmishing" phase. The aggrieved party sends a call over to the enemy, challenging them to present themselves at a definite battleground. This amounts to a formal declaration of war, and the enemy is allowed a day or two in which to make his preparations. These include clearing the battleground of undergrowth, a task in which both sides take part. If one side turns up while the other is engaged in this, it withdraws and waits until the other has finished before getting to work itself. On the evening before the battle, the shaman puts himself into a trance and informs the group's ancestors of what is about to happen, and thus secures their aid. A remarkable feature of the proceedings is a sweat ritual. The shaman takes each warrior's hand and wipes sweat from his armpit with it. Special magic is then used to bless the warriors' weapons, and other magical practices are intended to protect them and take away their fear.

For fighting, the Tsembaga begin by using only arrows and throwing spears. The arrows are unfeathered and seldom kill. According to Rappaport, this relatively innocuous skirmishing can be regarded as an attempt to limit bloodshed as much as possible; this phase lasts for several days, giving heated feelings a chance to calm down. On top of this, allies, having no direct interest in the conflict, try to exercise a moderating influence. During this phase, attempts at mediation are made by neutral third parties friendly to both sides. When the Dimbagai-Yimyagai and the Merkai clan of the Tsembaga fought, neutrals stood on a little hill overlooking the battleground, reminding both sides that it was wrong for brothers to fight; they called on them to leave the field and threw stones at them—a behavior pattern very like that of modern states.

During the skirmishing, the two sides eventually approach within hailing distance of each other. On the one hand, mutual denunciation and insults enable them to give free rein to their feelings without doing any physical harm to each other, while on the other hand, the possibility of reaching a peaceful settlement still remains open, as it does throughout the skirmishing phase. At the same time, the two sides have an opportunity of appraising each other, and one of them may realize that it would have no chance in a real fight and will therefore try to secure peace.

If no reconciliation is possible, the "real" battle takes place. In this, battle-axes and spears are used. This too is introduced by special rituals, all of which need not be described here. Special battle stones* normally kept in a net bag on the floor of a hut are now hung up on the center post. By this act, the group pledges itself to its ancestors and allies in return for their aid in the coming real ax battle; the group later releases itself from the pledge by special rituals. Also, two pigs are sacrificed and the spirits are called on.

The other side, who, if friendly relations existed with them, were previously described as brothers, now become "ax men," that is, they have been formally declared to be enemies. Henceforward, there must be no contact with them except in battle; talking to them, looking them in the face, and eating food grown by them is forbidden. (In this connection, one should recall what was said about the function of communication barriers.) There are also certain restrictions on social relations with the enemy's allies. Once the decision to fight a real war has been made, the opportunity for resolving the conflict quickly has passed.

In the course of preparatory rituals, the shamans give their clan the names of enemies whom it will be easy to kill and of members of the group who are in peril. They claim to have received this

* These are small stone pestles and mortars made by prehistoric settlers in the area, to which magical powers are attributed.

information from the spirits. An interesting aspect of this ritualization is that not very many enemy names are ever given; in other words, this is a hint that the group should be satisfied with the number of victims indicated. A rough "killing quota," in Rappaport's words, is set. At the same time, the shamans direct attention to very definite individuals on the enemy side.

There are a number of things that the warriors are forbidden to do. For instance, before the battle they eat highly salted bacon, but they are not allowed to drink. As a result they become very thirsty, which eventually forces them to stop fighting.

The "real" battle begins very cautiously. At first, both sides are roughly equal in strength. Each waits for the other to be left in the lurch by its allies, which may happen if the clash is prolonged, since the allies are not directly involved in the quarrel, after all. If one group finds it has numerical superiority, it takes advantage of this to make a massive attack. Sometimes a group that has been left in the lurch avoids battle, collects its wives, children, and pigs, and flees from its territory.

Warfare can last for many weeks. If one side kills a warrior belonging to the other, fighting is broken off to allow the requisite mourning and funeral rituals to take place. The warrior who did the killing himself must carry out certain rites. All this does not require more than two days, but the truce generally lasts for several days, and warriors use the opportunity to cultivate their neglected plots. Before resuming hostilities, both sides again sacrifice two pigs to the spirits. If a warrior is severely wounded, there is a truce of several days. These interruptions in the fighting tend to permit a more conciliatory atmosphere to develop and create the possibility of a resumption of negotiations.

Hostilities can be interrupted by an armistice. Sometimes one side is put to flight, and that is when most casualties occur, because the victors kill everyone they catch, including women and children. They lay waste the fields of the defeated, but they generally do not take immediate possession of their land.

Armistice rituals are carried out separately by the two belligerent parties. Sacrifices are made to the spirits, and cordylines are planted.* If a group is driven from its territory, it plants cordylines in the territory of the friendly group that grants it hospitality. This,

* While this ritual plant is growing, the group may not take part in an ax battle. If it wishes to do so, it must first uproot the cordylines.

in conjunction with the simultaneous sacrifice of a pig, amounts to an invitation to the spirits of their ancestors to leave their old home and come to the new. This is equivalent to a cession of territory, and the enemy is now able to take it over.

For the time being, however, the situation is only an armistice. The battle stones are still hanging, and a number of taboos remain in force. The armistice may last for years. During this period, both sides engage in pig breeding. When a group has bred enough pigs, it arranges a pig feast to which it invites its allies. The feast lasts for months, and a large number of rituals are observed. A start is made by erecting frontier posts. If a defeated enemy has not resumed possession of its territory, but has planted cordylines elsewhere, it is assumed that its ancestors have departed too, and frontier posts are planted in strange territory. Then a dancing place is prepared, and the cordylines that have been planted are uprooted. The real feast (*kaiko*) can then begin. The battle stones are put back on the floor of the hut, which means that social contact with the enemy can be resumed. The guests begin the feast by dancing, and the women and children who accompany them mingle with the spectators. Next the hosts dance, after the shamans have spoken magic to prevent them from being outdone by their guests. The magic is also intended to cause the girls among the guests to be impressed and to choose their partners from among the dancers.

When the feast is at its height, the group is released from various taboos, particularly those affecting social contact with the enemy and its allies. It will be remembered that such social contact is restricted by various bans as soon as an ax war has been declared by hanging up the battle stones. There is a ban on entering an enemy's hut, cooking over a common fire, and eating food that an enemy has grown. By this time, these taboos are felt to be irksome by both sides, linked as they are by numerous marriage ties. The feast ends with the warriors being publicly honored.

Hostilities can be resumed after the feast, but in the long interval between planting the cordylines and the pig feast, feelings have generally cooled, and as a result, there is a good prospect of peace. To establish peace, pigs are needed, which means that two or three years have to pass, as the pig feast exhausts the existing supply. For the conclusion of peace, the parties, accompanied by their wives and children, meet on the frontier. Pigs' livers are exchanged. Women are also exchanged, or are promised to former enemies.

The object is to exchange a woman for every dead man on the other side; this means establishing more ties of kinship. In fact, the higher the casualties have been, the greater the number of ties established; this represents a reasonable assurance that conflict will not again escalate into war very soon.

Throughout all this, like a red thread, runs a constant concern to resolve the conflict, to prevent its escalation, and finally, if fighting in earnest cannot be prevented, not to shut the door on the possibility of resuming contacts. Negotiations take place, and throughout the conflict, there are recurrent opportunities for making peace, first by refraining from ax fighting and engaging in ritualized battle only, then by the resumption of truces as cooling-off periods (and assuring that they will be protracted by the planting and growing of cordylines). Finally, the taboos that forbid social relations are felt to be increasingly irksome, since they keep relatives apart, and this again favors the conclusion of peace. When peace is finally restored, ties are reinforced by marriages in a number directly proportional to the number of those killed.

Mediation by Third Parties

A remarkable feature is the intervention of third parties who condemn the fighting as a bad thing. They have no authority: they can use only persuasion and cannot enforce a cessation of the fighting. Apparently, that is one of the reasons that the resolution of conflicts between different groups is often so difficult. Within the group, both in animals and man, peace is ensured by authority through the rank system and the intervention of seniors—in modern states, this is a function of a special organ (the police)—but in intergroup relations, such developments are only embryonic.

We know from the Mount Hagen tribes of New Guinea that the escalation of a conflict can be successfully prevented by third-party intervention. Owa, a man of the Tika tribe, committed misconduct with the wife of another member of the tribe, who thereupon misconducted himself with Owa's wife. Soon afterward, the clans were about to engage in battle when some other chieftains intervened between them, holding their spears crossed. In such a situation, no one dares to attack; instead, they have to negotiate, and this case, reparation in the form of pigs was agreed on.[13]

Chieftains who are successful in resolving conflicts are highly respected for this ability. Strathern quotes what a Mount Hagen tribesman had to say on the subject:

> "Nimb was a major big-man of my clan, and he married five wives. He was a great peace-maker. The reason why he was able to make peace was that he made all men afraid. . . . When all the Mokei clans fought against the Yamka, he forbade them to carry on, and they stopped. He protected the Mundika tribe and the Nengka Kwipanggil. [Some 1,500 persons!] He used to present shells and pigs to those threatening fights, telling them not to quarrel, since they were all sisters' sons and brothers of each other. He broke up fights just when they were starting."[14]

"Big-men" were respected above all for their ability as peacemakers. If a war had already caused deaths, parity had to be established before the parties could be persuaded to make peace. No money payment was acceptable as retribution for a death, and only when there were an equal number of dead on both sides did they say: "Now we are quits. Now we can look each other in the face."[15] The Australian government enforced the system of reparation in the form of pigs as the only means of repayment, and this has since been recognized as a good thing by the Mount Hagen tribes. Now, Strathern reports, they acknowledge: "Before we fought and killed each other, and this was bad; now a good time has come, and we can pay for killings. . . ." *Moka* feasts, with which we shall deal later, play a big part in such settlements. The wish for peace evident in the funeral speeches quoted earlier reappears in this statement.

Formal Peacemaking

It is worth remarking that even at a very early stage of development, a formal conclusion of peace is expected. It closes a chapter and at the same time builds for the future. To those who seek to understand the laws underlying cultural rituals, the patterns by which primitives make peace are revealing. Many of the rituals are based on innate dispositions. This applies to the ritual of exchanging gifts and eating together. In addition, purely cultural mechanisms developed with the exchange of marriage partners.

If a group wants to end a war, the first thing it must do is communicate this to the enemy. There are various ways of doing this. The Jalé of West Irian address standarized verses to the enemy. Koch reports that a party seeking peace sang the following again and again:

> Fighting is a bad thing, so is war.
> Like the trees we will stand together,
> Like the trees at Fungfung
> Like the trees at Jelen.*

At the same time the Jalé tried to shift responsibility to a scapegoat:

> Weli,† it is your fault!
> Weli, it is your blame!
> That fire ate the River's water,‡
>
> That fire ate the Sévé River,
> That fire ate the Jaxólé River,
> Weli, it is your fault!
> Weli, it is your blame![17]

The Kiwai-Papua inform the enemy of their wish for peace by laying a branch over the path to the enemy village. If the idea is acceptable to the enemy, they too lay a branch over the path; if it is not, they turn the enemy's branch so that it points in the latter's direction. They also lay small sticks on the path; the number of sticks shows how many enemies they want to kill before they will be willing to make peace.

If the idea of peace is acceptable, a number of men accompanied by their wives make their way to the enemy village. The women walk a few paces ahead. It is taken for granted that bringing their wives is a demonstration of peaceful intentions, and their reception is certainly friendly, at any rate on this occasion. The men break their knives as a sign of peace and exchange arm bands. During the night, the hosts sleep with the visitors' wives—a practice known as "putting out the fire." This is followed by a return visit on which the same thing happens. Everyone drinks together and hostilities are declared to be ended. Girls are married to close relatives of the dead, as a means of compensation.[18]

The Mount Hagen tribe makes peace on similar lines. When parity in the number of dead has been reached, hostilities gradually cease

* Fungfung and Jelen are places on the mountain ridges between the valleys in which the belligerents live.

† The man who began the war.—I. E.-E.

‡ I.e., that many men and pigs were lost in the war.—I. E.-E.

and a decision to make peace can be made. Negotiations take place, pig meat is exchanged, and vows are made to keep the peace. The two sides face each other with crossed spears with intermediaries between them, also with crossed spears. When the two sides have made their peace vows, they both sit down on the ground. The vow itself consists of the two sides reciting the following verse, one side repeating the line just spoken by the other:

> The birds Towa and Kopetla shall again leave their footprints behind.
> Women and pigs shall again go backward and forward between us.
> The trampled grass will rise again and grow over everything.
> So shall we again have peaceful relations with each other.
> We shall live in peace and multiply.
> We shall make no more war on each other.

Everyone then sits down and negotiations begin. Meat is exchanged, the war is discussed, and agreement is reached about reciprocal reparations. In spite of the agreement, a certain amount of mistrust remains between the parties, but women begin paying visits, and if nothing happens to them, men pluck up courage and do the same. When the dead have been finally paid for, it is considered that peace has been established.[19] Similar patterns of peacemaking, in which the speakers of both parties refer to the many enterprises they had engaged in before the war broke out, were recently observed among the Eipo in West Irian.[20]

Such peace feasts are sometimes marked by ritualized fighting. Helena Valero graphically describes what happened when two enemy Yanomami tribes visited each other with a view to a reconciliation. After they had drunk banana milk and sniffed an intoxicating powder together, the hosts said: "You are excited, we are excited, we must calm ourselves," and a ritual exchange of blows then began. Men took turns striking each other's chest muscles with their fists. Then, with long hardwood staves, they struck each other alternately on their clean-shaven heads. Before doing this, they said: "I sent for you to see whether you are really a man. If you are a man, we shall now see whether we shall soon become friends and whether our anger is passing." To which the reply was: "Speak to me like that, go on speaking to me like that, strike me, we shall be friends again." Many beat each other into a state of insensibility. Finally they said: "We have struck you some good blows, and you have struck us some good blows. Our blood has flowed, and we made your blood flow. I am no longer excited, our anger has subsided."[21]

What happens at peace celebrations in the Andaman Islands is even more ritualized. The men of the forgiving party dance into the village of their former enemies. They make threatening gestures at the men of the village, who are quietly awaiting them. Eventually, the leader of the dancers seizes one of the village men and shakes him, jumping up and down wildly. Then he goes on to the next man and does the same to him, while the man he shook first is then shaken by the man behind him. This goes on until every dancer has twice shaken each of his former enemies, once from the front and once from the back. Then it is the women's turn to shake each male member of the enemy group. When this has been completed, the two groups weep together. They remain together for several days, hunt and dance together, and exchange gifts. Peace has then been sealed.[22]

Means of Maintaining Peace and Avoiding Conflict

In connection with the reconciliation feasts of the Mount Hagen tribes, gift-exchange (*moka*) rituals take place at which one group gives the other pearl mussels and pigs. This creates an obligation; after a certain interval, a return gift of at least equal value is expected, and this again is handed over as part of a large-scale celebration. The group making the return gift generally tries to gain prestige by giving back more than it received, which results in a kind of competition.[23] A number of groups often take part in a *moka* ceremony, group A giving to group B, B to C, C to D, and so on, until the last link in the chain has been reached. After an interval, each donor group receives a return gift from the recipient, and the *moka* chain has then been completed.

On the occasion of one such *moka* ceremony, dancing took place and speeches were made. We recorded one of these speeches near Mount Hagen in August 1973, and the tape was translated by Martin Wimb. I quote the speech of the "big man" who organized the occasion:

> "At the Maninga and Keli fights many men died. Because of those fights many ceremonies took place, and this is the last of them. The ceremonies now come to an end in the presence of some old men

who took part in those fights, and I hope that those old men, who are still with us, will be glad. Those old men, who once had many pigs and other wealth, are now very old. The men of the present day are men of alcohol [beer],* and they no longer have as much wealth as those old men.

"I was the one who caused the fights of the Kopi, Nokaba, Maninga, and Yimi tribes, and that is why I gave pigs to all those tribes, and last of all I have given them here to the Kopi and Nokaba tribe, and this is the last time. Some old men who were at the fights are present. So the troublemaker, as most of the tribes call me, has rid himself of his oppressive guilt, because I have now satisfied all the tribes that lost men in the fights caused by me. I can now think of no tribe that has been left out, and from now on I shall be a free and friendly man for all time."

Moka ceremonies undoubtedly serve the cause of peace. At the same time, they contain an element of competition for dominance. The ceremonial exchange of objects of value reinforces alliances between parties who regard each other with a certain amount of mistrust, it is true—but as allies, not as traditional enemies. By outdoing the other party's gift, one gains prestige.

The resolution and prevention of conflict by binding rituals such as feasts follows basically the same pattern in many cultures. In this connection, the reader is referred to my description of the palm-fruit festival of the Yanomami Indians.[24]

The tribes of Central Australia have found yet another way of avoiding territorial conflicts—by mythical ties to localities. They believe that each group is descended from a totem animal that gave them the territory in which they live, and they honor certain striking features of the landscape as traces left by these totemic ancestors. The men of the territory may enter these sacred places only for the purpose of holding special ceremonies, and it is only these places that they defend against intruders, which is seldom necessary. Every adult male possesses a sacred board on which are recorded, in highly symbolic form, the wanderings of the sacred ancestral totem animal from which the group is believed to have descended and the important landmarks in the area connected with the myth. The engravings are the armorial bearings, as it were, both of the territory and of the individual. Their strong emotional bond to their territory prevents the inhabitants from setting out to conquer other territory, since they would not feel at ease there, for it would be the home of the

* Alcoholism came into the country with the Europeans.—I. E.-E.

spirits of other men's ancestors. This idea has, as I mentioned, also been developed by the Tsembaga, who will not occupy strange territory until its previous occupants have planted cordyline trees elsewhere and thus summoned their ancestors to their new home.

As a further measure, these Australian aborigines have developed an interesting division of functions. Each territorial group promotes the welfare of its totem animal by rituals at its holy places, and not only within its own boundaries. The honey-ant clan sees to it that there are always enough honey ants; the emu clan looks after the well-being of the emu; and the kangaroo clan is responsible for the kangaroo. Thus each group is important to all the others, which also has an inhibitory effect on destructive warfare. Conflicts are not prevented entirely, however. There are quarrels about women, but these are often—though not always—settled by nonlethal dueling.

Their mythical bonds to places assure these Australian aborigines of their territory without entailing any special need to defend it. As a result, a group can be more open in relation to its neighbors; evidence of this is provided by Hiatt's accounts of the Gidjingali of Arnhem Land, in Australia. Each clan has its own territory, but only its holy places are jealously guarded. The rest of the land is also the property of a group, and it has an emotional tie to it, but the various clans visit each other and gather food in each other's territories when they do so. Group A, for instance, which generally lives on the west side of the Blyth River, regularly visits group B on the other side of the river in August or thereabouts. The nuts of the cycas palm, which are highly valued and ripen at about that time, are found there only. A month later, both groups go together to hunt geese in some inland swamps belonging to groups C, D, and E. In this way, the whole area is much more effectively exploited.[25]

PATTERNS OF CONFLICT CONTROL

Anthropological literature contains many detailed descriptions of conflicts and their resolution. They all follow a very similar pattern, showing that at a very early cultural level, men sought to settle conflicts without bloodshed. They did so by means of rules that, although culturally developed, harked back to their existing heritage. Every advance in weapon technique and every new social development necessitates a readaptation of cultural control conventions. In the late Middle Ages, for instance, the ceremonial of knightly combat was made obsolete by the invention of firearms. The new weapons, which completely violated the rules of chivalry, were condemned, and the knights wanted them abolished. To them, war was a game that took place in accordance with a precise code of honor. Firearms, with which anyone could start a fight by treacherously opening fire from behind cover, totally ignoring the code of chivalry that had taken so much time and trouble to develop, seemed utterly repugnant. Luther thundered against these new-fangled cannon and muskets, which he described as inventions of the devil, since neither strength nor courage availed against them.

Firearms are still murderous weapons. Man has succeeded only in adapting to them strategically, without ritualizing the clash itself, though there are agreements not to use certain types of ammunition, such as dumdum bullets, that cause excessively gruesome wounds.

Our age has provided us with terrible weapons: explosive and incendiary bombs, poison gas, bacteriological weapons, and finally, nuclear explosives. Because of their fearful effect agreement has been reached not to use some of them, or at any rate, an effort is being made to reach such an agreement. But the road to it is thorny—in part because economic interests are an obstacle. According to the *Süddeutsche Zeitung* of August 22, 1972, a protest by a Dr. Theodore Tapper against the use of napalm "bounced off a wall of moral ignorance and greed for material gain," at the annual general meeting of the Dow Chemical Company. Dr. Tapper had called on the shareholders to refuse their consent to the further production of napalm, and showed slides of terribly mutilated victims in the Vietnam war. Napalm, as we know, ignites on impact and cannot be extinguished. It sticks to everything and develops a temperature of 2000 degrees centegrade. Carl Gerstacker, the chairman of the company, rejected the protest and was applauded when he read a letter from South Vietnam in which an American soldier expressed his appreciation of the product. The International Committee of the Red Cross has tried, so far in vain, to have it banned. In November 1972, the Political Committee of the United Nations screwed itself up to passing a resolution "deploring" the use of napalm in armed conflicts.

New techniques raise new problems in the ritualization of conflict. Partisans, for instance, so long as they wear a uniform, can be recognized as taking part in a war of liberation, but surreptitious operations by civilians are considered to infringe the rules of fairness, because they involve the civil population in a clash, when it would be preferable to keep them out of it in order to limit suffering. Also, we are helpless at the moment against the latest manifestations of terrorism, and one almost has the impression that effective conventions against it are being developed only on a trial-and-error basis rather than through rational planning.

The ritualizations that have taken place so far enable one to conclude that there is a trend toward defusing intergroup conflict although the functional conflict between aggression and friendly bonding has not been resolved; this is a point to which I shall return. These ritualizations have not advanced as far as they have in intragroup relations, in which aggression is effectively kept under control. Behavior patterns available in the form of phylogenetic adaptations are involved as well as others that have been developed, culturally. Man is trying to defuse intergroup conflicts on lines similar to those successfully used in intragroup conflict.

Among the higher vertebrates, interspecific aggression is inhibited by the mechanism of fear; as a result, only in the case of desperate need does a predator approach a prey capable of defending itself, and the prey, in turn, attacks only if the critical distance has been crossed. When they attack, they generally aim at killing. In intraspecific conflict, they do not do this. Among some higher mammals, however, members of the group are treated differently from strangers, and in a number of species destruction is the goal in intergroup conflicts, while in intragroup conflicts, conspecifics are spared.

Patterns of Conflict Control in Animals and Men

The following pages show how destructive conflicts are avoided in intraspecific relations in the animal kingdom and in intragroup and intergroup relations among men.

ANIMALS

1. *Ritualization of the clash.* Fighting is preceded by a threat duel; the fight itself is nonlethal. In extreme instances, no physical trial of strength develops. We know from comparative studies that this nonlethal dueling has generally developed from serious fighting.

2. *Submission postures* enable the loser to end the fight. In doing so, the animal switches off fight-releasing signals. In accordance with the principle of antithesis, the animal's behavior is the opposite of intimidation behavior—making itself smaller, for instance. Among higher vertebrates, appeals for contact are made, especially in the form of infantile signals.

3. *Intervention by third parties* occurs among higher mammals. High-ranking animals intimidate quarrelers by threats. In some species, they also pacify by "friendly" rituals (greeting). This brings about a change of mood; that is, drive systems antagonistic to aggression are activated.

MEN

1. *Ritualization of the clash* in essence resembles its counterpart in the animal kingdom, but among humans is both phylogenetic and cultural. Phylogenetic ritualizations determine expressive behavior (threatening) and play a large part in intragroup conflict in particular. Cultural rules control handling of weapons. This applies in both intragroup and intergroup conflict, but in intergroup conflict (war), ritualizations are less advanced. Verbalized aggression (for instance, song duels, p. 96) can take the place of physical trials of strength.

2. *Behavior patterns of submission* are partly innate (pouting, weeping), partly culturally developed (laying down arms). Appeals can be verbalized. Breaking off contact ("cutting off") by turning away, temporarily leaving the group, or refusing to talk are obviously aggressive, but they are in addition well suited to prevent conflict by simultaneous distancing.

3. *Intervention and mediation by third parties* occurs in both intragroup and intergroup conflict. Appeals consist of references, on the one hand, to an existing or fictitious relationship (the moral appeal: "It is not right for brothers to fight") or, on the other hand, to the terrible consequences; finally, threats are used. Among Bushman children, the aggressor is punished by older children, and the victim of aggression is consoled. The intervention of a mediating authority plays a big part in the resolution of intragroup conflicts, but a smaller part in intergroup disputes, though in modern times, there has been an increase in this direction.

We see culturally newly developed forms of resolving conflicts by third parties at work when a group or a member of a group is authorized to use his good offices to bring about a settlement or to pronounce judgment based either on recognized usage or codified law. Henry Kissinger's efforts on behalf of the United States to bring peace to the Middle East, as well as the United Nations efforts to act as a mediator between warring parties, are recent examples. In intragroup relations, judicial authorities and specially instituted guardians of order (police) exercise effective control of aggression. They are also empowered to penalize breaches of the peace by special sanctions. There are the beginnings of a development in this direction in intergroup (interstate) relations, but law and the executive power to enforce it have not been sufficiently developed. The mere fact that, at the interstate level, the right to self-defense is basically recognized shows that, from the evolutionary point of view, interstate relations are still governed by "primitive" law.

ANIMALS

4. *The formation of a social ranking order* makes the behavior of a member of a group in its social context predictable for every member of the group. This certainly prevents a great deal of friction. It has sometimes been claimed that aggressions are handed down the rank pyramid and thus either thinned out or—exactly the opposite—concentrated on a scapegoat, who thus acts as a kind of lightning rod. But these suggestions are not based on any definite evidence.

5. *Behavior patterns indicating readiness for friendly contact.* Social mammals have a number of these at their disposal. Most derive from the repertoire of mother-child signals, some derive from female sexual signals. They have a pacifying effect and establish or consolidate the friendly bond between members of a group.

6. *Avoiding provocation,* for instance, by concealing fight-releasing signals in the group, is another means of preventing conflicts.

7. *Development of rules* that, for instance, prevent disturbance of a bond between two members of a group by a third party, as by kidnaping or seducing a marriage partner or child.[27]

8. *Evasive action* (activation of the flight system) can be utilized to avoid conflict.

MEN

4. *The role of ranking order* among men is similar to its role among the social mammals. Aggression is often diverted to scapegoats or minorities, which reinforces the cohesion of the group.

5. *Bonding and conciliatory patterns of behavior* are inborn in man and, like those of the higher mammals, are derived mostly from mother-child behavior patterns. Cultural ritualizations appear in the verbalization of appeals, which, however, remain basically the same: they are verbalized infantilisms and care behavior. Verbal gifts, for instance, are given in the form of good wishes. Greeting rituals on meeting and at a feast show that cultural ritualization and phylogenetic adaptation are structurally geared in with each other. Greeting rituals and feasts are constructed according to universal rules, and their appeals are everywhere the same, though they are expressed in different forms. The *exchange of gifts* plays a big role both in intra- and intergroup relations. In relations with other groups, it expresses a desire for peace, and several times, it has been pointed out that trade was originally an exchange of goods in the service of bonding.[26] This can still be observed among tribes that engage in trade without having any need to do so, since everyone has the same resources and can readily obtain the goods concerned for himself.

6. *Avoiding provocation* by eliminating aggression-releasing signals can be achieved by, for example, giving up weapons, which are generally no longer carried by civilians in everyday life. A leveling out of wealth by a distribution of property also plays a part in internal pacification both among primitives and in advanced civilizations.

7. *Legal limitation of aggression.* The establishment of a legal authority to settle disputes presupposes the development of a body of law that taboos aggression. The escalation of conflicts within groups is prevented in this way, and property and territorial integrity are guaranteed.

8. *"Cut off"* and *evasion* play an important role in avoidance of conflict among men, too.

ANIMALS

These patterns of reaction are present in animals essentially as phylogenetic adaptations, though animals learn which patterns of behavior to use against which partners, depending on whether the latter are higher or lower in the ranking order, for instance.

I know nothing in animals corresponding to the patterns of conflict control described in points 9 through 12 on the opposite page.

MEN

9. *Safety-valve mechanisms* enable aggressions to be acted out without causing harm to the group.

10. *Attempts are made to prevent intergroup conflicts* by the establishment of *marriage ties* spanning boundary lines. This is one of several ways of keeping channels of communication open and combining groups into larger units.

11. *Keeping possibilities of contact open* between hostile parties, through feasts, for instance, or by permitting certain persons to move freely as mediators between them (among the Yanomami Indians, for example, old women can move unhindered between hostile groups). Institutions such as the Red Cross or the United Nations fulfill a similar function at a more advanced level of civilization.

12. *Rousing humanitarian feelings* and *political and economic integration* are other ways of preventing conflict between groups. Use is made of values developed in the family (brotherliness); in other words, the family ethos is extended. Included under this heading is education for peace, with a conscious rejection of aggressive models.

Of all the forms of behavior listed above, the following are typically human: conscious education for peace, development of civil and international law, integrating bigger groups by an extension of the family ethos, and avoidance of conflict by the exchange of marriage partners between groups.

The other ways of resolving conflict also appear in the animal kingdom, but they are given their specifically human stamp by cultural ritualizations—through verbalization, for instance. That applies to ritualized combat as well as to the large number of usages that consolidate a bond, resolve tensions, promote the solidarity of a group, keep channels of communication open between enemies, and finally reconcile belligerents and cause them to make peace.

MODELS OF HARMONIZATION AND EDUCATION FOR PEACE

> In the Aisne Valley, it was often misty in the morning in the autumn. Since there were a great many men who shot but none who shot game, a rich variety of birds and beasts of the chase sprang up between the barbed-wire entanglements. . . . Our major . . . was very fond of partridge, and he sent for his shooting equipment and on such misty mornings went out stalking between the barbed-wire entanglements. Instead of his officer's cap, he wore a hunter's hat, to the great delight of his men, and he also carried a hunting stick, and he slung his gun and gamebags over his shoulder. In this getup, it was evident that he was out shooting game.
>
> To us, these shooting expeditions were a delicate matter. . . . We did not want to lose our commanding officer. So we secretly set up two machine guns so as to be able to give him covering fire in case a gust of wind suddenly blew the mist away. . . . One morning the disaster occurred. The wind began to blow and the mist vanished in seconds, just when the major was standing right in front of the French barbed wire, only a few yards from the enemy trenches. The obvious thing would have been for him to have flung himself on the ground and tried to creep back to us, using the cover of the tall grass and the rather undulating terrain. To our amazement he did not do so. We forgot to fire.
>
> The French must have been even more taken aback than we were. They cautiously peeped over their parapet, and nobody fired. One man may have raised his rifle, but the major threatened him with his hunting stick as if he were Frederick the Great. Then ten or a dozen poilus suddenly broke cover, laughing and shouting: "Bonne chasse, colonel! Bonne chasse!" The major gave them a friendly wave and walked slowly back to our trenches with a hare over his shoulder. If anyone had shot him during those few minutes during which he offered a magnificent target, French and Germans alike would have felt it to be murder. Five minutes later killing someone on the other side was no longer murder. What is the secret behind that paradox? (Peter Bamm, *Eines Menschen Zeit* [One Man's Time]).[28]

In earlier chapters, I have discussed the question asked by Bamm, and I can answer it. This paradox in human behavior is the result of the existence of two norm filters that bid us do different things and are therefore in conflict. If we dissociate ourselves from the enemy, we shall be inclined to kill him in accordance with the cultural norm filter. But as soon as we establish personal contact with him, our innate inhibitions against aggression begin to affect us, and behavior patterns of making friendly contact are activated also. Bamm's story, which well illustrates this, could be supplemented by many others. It is well known that in positional warfare, communication between combatants has to be forbidden; otherwise the troops begin exchanging cigarettes across no-man's-land and what is known as "demoralization" sets in. The degree of ease or difficulty with which the biological norm filter makes itself felt in individual cases depends on a multitude of circumstances, of course. Cultural indoctrination, fear, readiness to obey authority, all these things and others play important roles, but the tendency is there, even if it does not become effective. It is certainly not a sufficient basis for peaceful human coexistence, but it is a vital condition for it.

At the root of men's desire for peace is the pressure to bring the cultural filter into harmony with the biological filter. However, the universality of the desire is disputed by many, including Wintsch, who maintains the opposite: "Peace, interpreted in the traditional sense as a legally ordered and peaceful form of human coexistence, does not seem to be a general need. There is hardly such a thing as a 'peace drive' that can be isolated from among the many determinants of human behavior in the way in which the feeding drive or the sex drive can be isolated. . . . I actually believe . . . we are entitled to say that what man wants is not peace, but appeasement of his needs."[29]

The last sentence solves the apparent conflict between Wintsch's views and mine. One of man's needs is life in harmony with the biological norm filter. If biological and cultural norm filters are in disharmony, appetences are activated until the discrepancy is removed. Our innate norm filter lays down how we are to react to certain appeals by our conspecifics; among these, certain submissive appeals set in motion inhibitions on aggression.

I must refer here to an idea expressed by Mitscherlich, who believes that basically we are afraid of peace,

> certainly in the deeper, hidden layers of our mental organization, which of course also contains within itself the great experiences of the history of the development of the species. The feeling of being robbed of the possibility of expressing collective aggression is unconsciously regarded as a highly dangerous, defenseless condition; this is reflected in the vague displeasure connected with associating oneself with peace in a more than rhetorical fashion, and may be

> one of the reasons why the term world peace sounds so hollow and dishonest in many mouths.[30]

The truth of this is that fear based on mistrust has hitherto prevented general disarmament, but that does not prove that man does not want peace. He also wants his security to be guaranteed. We as individuals feel more secure in the modern states of today with their legal systems than men did in the days of club law, and we are less anxious, though we go about our business unarmed.

When we consider the prospects of lasting peace, we must first of all ask two questions:

1. Is our motivational structure equipped for peaceful coexistence in modern mass societies?

2. Can the functions of war be performed without war?

On the first question, Lorenz has expressed a rather pessimistic view. He believes men are adapted for life in small communities. Our intelligence is perfectly capable of understanding the principle of loving those unknown to us, but emotionally we give no backing to it. One can love only those one knows. The principle of loving people of no matter what origin can be grasped by the reason but not by the heart, and with the best will in the world, that cannot be altered. On this point, according to Lorenz, we men are only on the way to becoming men; we are, as it were, the "missing link," "not quite good enough for the demands of modern social life," and only those great forces that effect change in species—mutation and selection—could achieve any modification of this. Lorenz hopes that reason is practicing rational selection.

Von Holst takes a similar line. He says that a peaceful race of men must first be bred. He believes that our inclination to intolerance is so great that if an ideology propagating peace one day triumphantly spread all over the world, it would promptly split into two parties—one representing the true and the other the "heretical" doctrine.

The reader will have concluded from all that I have said that I view the situation optimistically. I regard man as preadapted to life in mass societies and thus as "good enough." His innate ethical code that tells him not to kill is supplemented by a strong pressure to make contact with those unknown to him, in other words, to fraternize. The behavior patterns of bond formation and the urge to bond are, as we have said, based on phylogenetic heritage, and thus the commandment "Love thy neighbor as thyself" has a biological foundation.

The astonishing success of doctrines that advocate love of one's neighbor is based on this. Man does not want peace contingent on avoidance of contact with strangers. Strangers alarm him, but nevertheless, overcoming his alarm, he seeks contact and is prepared in principle to see all men as brothers. As the history of national states plainly shows, it is perfectly possible by way of symbols and common tasks for men to identify with others whom they do not know, and to do so with strong feelings. Hitherto, such "fraternization" in anonymous mass societies has generally taken place under the influence of a real or simulated threat. But that does not prove that an enemy dummy is always required to bring about fraternization in an anonymous society. Men have united in a common struggle against the forces of nature just as they have in fighting "enemies." There is an abundance of tasks that unite humanity, and there will be no lack of such challenges in the future.

The idea that an enemy is needed to bring about the unification of a group is based on a fallacy. The fact that it is usually an enemy that unites a group leads to the assumption that an enemy is necessary for this purpose. On this, Mead comments: "It is frequently argued that federations have been dependent upon external enemies and therefore are impractical forms to use as models for a single world order. But internal order has been attained by societies without the need of invoking external enemies, once the social forms made it possible to identify all members as equally human and exempt from treatment as non-human. Slaves have been freed, children released from the brutal domination of parents and teachers, women given rights, and minorities given full citizenship, as the social identifications have been altered, in belief and in practice."[31] And in *The Origin of Species* (1859), Darwin said it was men's duty to extend their feelings of sympathy from the members of a small group to all humanity: "If man advances in civilization and smaller tribes are united into bigger communities, the simplest consideration will tell the individual that he must extend his social instincts and sympathies to all the members of the same nation, even if they are personally unknown to him. When this point has once been reached, only an artificial boundary remains that keeps him from extending his sympathies to all men of all nations and races." Finally, Gehlen points out that this is an extension of the family ethos: "The ethos of loving one's neighbor is the family ethos. It first springs to life within the extended family, but it is capable of extension until it embraces the whole of humanity."[32] This is consistent with our view that it is with the aid of modified mother-child signals that we establish and maintain friendly contact with our fellow men (p. 101).

Man has a sense of family, and is in a position by way of symbolic identification to regard humanity as a family. The child as unifying symbol* would well express a major concern of us all, that is, the future of our children. The answer to the question whether our motivational structure makes us capable of peace is yes.

The second question, whether the functions of war can be performed without war, is closely connected with the question whether war does not still bring benefits in the present age. Are the biological norm filters perhaps the only forces urging us in the direction of peace while selection continues to drive cultural development in the opposite direction?

In the past, wars have certainly brought benefits to the winners; cultural pseudospeciation drove on human evolution and bestowed on us a colorful multiplicity of cultures. But now, even the remotest corners of the earth are populated, and the wars of recent centuries have begun to destroy the multiplicity of cultures we value so highly. The defeated who are unable to retreat are destroyed. After a chapter of human history in which selective conditions worked toward a ritualization of conflicts, we are now confronted with the appalling possibility of a reversal of that trend. With the brutalization of conflict and the death of cultures—particularly primitive cultures—a process of large-scale dedifferentiation has set in, a process of involution that threatens the variety of human cultures and in the long term restricts the evolutionary chances of the human race.

Moreover, in the present state of weapon technology, it is improbable that there would be any winner in a war conducted with nonconventional weapons, and with further technological advances, the possibility can only diminish further.† Today a nuclear war would most probably lead to the end of civilization, and tomorrow it might end with the annihilation of mankind. In evolution hitherto, fighting and killing have led to improvements in weapon technology. But apart from the fact that this process is repugnant to our moral sensi-

* This idea comes from Fremont-Smith (see I. Eibl-Eibesfeldt, *Ethology: The Biology of Behavior* [New York, London, 1975]).

† This stalemate is, however, extremely unstable and dangerous, because a side possessing only a short-term advantage in defensive technique (for instance, in the development of a completely effective rocket-defense system) would actually be forced to attack.

bility, today it carries an intolerable risk. It must therefore be succeeded by a system based on rational planning.

Such a decision for peace corresponds to our inclinations. It presupposes recognition of the functions of war and the performance of those functions by peaceful means. Those who regard war only as a pathological deviation are bound to go astray in their search for a cure, because the idea will never occur to them of performing in a different way the functions previously performed by war.

I have already mentioned that this involves securing the foundations of the living conditions of the different peoples. Until now, nations have acquired their territories by conquest, and defended them and used force to obtain the raw materials they needed. How the problems that have hitherto been solved by war would be differently solved under a peaceful world order is a question that lies outside my competence; politicians and economists must find the way. The biologist, with his knowledge of different paths of development, the general laws of evolution, ecology, and human nature, can only stand by and advise and point out the possibilities.

We know—and have known since Malthus wrote *On Population* (1798)—that the stream-of-life principle (p. 22) tends to a ruthless multiplication of its biomass and exploits every passing opportunity. In years of favorable feeding conditions, field mice and lemmings multiply inordinately. They do not know that years of less generous food supply will follow, and from the point of view of preservation of the species, that is not important. The inevitable population collapse after the population explosion leads to migration on the chance of finding new country to settle in, and to wholesale death. But there are always some survivors, and after rigorous selection, they carry evolution further. The stream of life is not obstructed by wholesale death; on the contrary, while it overflows its banks, floods and lays waste wide territories, it also finds new tributaries and channels to follow. Pressure of population enforces adaptation to ecological niches that are still free.

Man is in the midst of a phase of mass multiplication, but in his case, the individual counts; so the prospects of further collapses of population of the kind that took place in the Sahel area and in Bangladesh should alarm us greatly, since in the future, they are likely to occur on a bigger scale. The industrialization of Europe that led to a population explosion is totally dependent on the use of fossil fuels, of which there are only limited supplies. When they have been used up, nuclear power may bridge the gap for a while, but fissionable elements are rare too, and the problems of nuclear fusion under technically controlled conditions have not yet been solved. If they are not solved in time, Europe may be faced with a population collapse that it will be able to escape only by a flight forward, i.e., through war. That is how evolution has advanced in the past. Population pressure brought about by opportunist multiplication can lead to the conquest of new territory, to expansion, or to collapse. The consumer society accords with the principle of the uninhibitedly multiplying and consuming stream of life. This is a high-risk development, and we know that in the course of evolution, many species have fallen by the wayside. Also, such a method of evolution is not in accordance with our ethical ideas. So we must plan differently. Man must master the stream-of-life principle.

Malthus pointed out that mankind doubles its biological potential every twenty-five years, but that our food resources grow more slowly, and that the inevitable consequence is poverty and hunger. Until recently, Malthus was laughed at, but now our laughter has died. But insight into the necessity of worldwide birth control makes headway very slowly. There is no real talk of limiting population, though our planet's capacity is now known quite accurately. Meanwhile, we human beings have not yet decided for peace. In spite of the lip service they zealously pay to it, the power groups still strive to outdo each other with the aid of more power, and that also means more men.

There are various models for a life of peace. One can conceive of a viable civilization in which people were completely manipulated and felt completely free and happy in the monotonous repetition of indoctrinated credos. Aldous Huxley offers us such a model as a warning in his *Brave New World.* On the other hand, Skinner in a certain sense sees it as an ideal worth aiming at.

Lorenz asks why life in such a manipulated system, offering absolute security and stability based on a system of well-thought-out doctrine, does not seem desirable to us. He concludes that our objection to it is not rational but is based on a value judgment, for we feel values without being able to give reasons for them. We talk of higher and lower animals, and the standard by which we judge the level of their development is not the degree of their adaptation but, according to Lorenz, the amount of information included in their blueprint; and not just the actual amount, but also the potential to acquire further information, and consequently, the ability to learn and behave intelligently—in short all the things that go to make up individual adaptability.[33]

Included in this, of course, is the dynamism of an organism coping actively and exploratively with its environment, which we put higher on the scale of values than passiveness. On the other hand, we put a negative value on developments leading to a loss of differentiation, even though a perfectly adapted organism is the outcome. When the highly differentiated barnacle larva *Sacculina,* possessing eyes and other sense organs, a nervous system, and an apparatus endowing it with mobility, turns into an eyeless parasite incapable of movement that permeates the body of a host like a fungus, it fills us with horror. From the evolutionary viewpoint, such specialization represents extreme perfection in one direction and, at the same time, a sacrifice of evolutionary prospects, for it is difficult for an organism adapted with such extreme lopsidedness to change and readapt itself when called for by environmental changes. Its chances of maintaining itself in the stream of life are slight. But that seems to me to provide a basis for our valuation. Man as a species has certainly undergone selection for universality and differentiation. Our values provide us with a protection against involutionary developments; involution is implicit in the model of the totally manipulated society that would involve restriction of creative liberty. A forced conformity of cultures would cut back man's evolutionary prospects.

Herein lies the greatest danger to any evolution planned and guided by us. As soon as we direct it to a definite goal, we run the risk of narrowing the spectrum of possibilities and thus setting in train a process of involution. Differentiation, manysidedness, and openness to the world are human properties that must be retained. The evolutionary possibilities of man lie in the colorful multiplicity of cultures and peoples. Every culture is an experiment that opens up new possibilities and thus increases the chances of our species' maintaining itself in the stream of life; and that multiplicity must be preserved.

If we decide for peace, a system of international law will have to be established that will be as detailed as the civil law of national states and will take the place of the primitive law of self-defense that has hitherto prevailed in international relations. This system presupposes judicial agencies having the power to impose their decisions. An international police force will have to be set up, equipped with modern conventional weapons and composed of representatives of all countries. With such a police force in being, the great powers could begin gradually dismantling their arsenals of weapons of mass destruction.

An international agency of this kind would not only have to guarantee the integrity of the various national territories; it would also have to ensure that reserve territories were kept available so that, in case of need, populations whose homeland had for any reason become uninhabitable could take refuge in them. Supplies of the major raw materials would have to be internationalized and fairly distributed. World trade would have to develop more along the lines of a cooperative distribution system, based on trade-offs of productivity. Finally, birth control would ensure that nations were not forced into a flight forward by overpopulation.

The development of the United Nations has so far been toward coexistence among cooperating but independent nations recognizing a common charter. It is also conceivable that a stable world order might result from a world power's subjugating all others and establishing a world government itself. At the present time, it is improbable that a single power would succeed in doing this by the force of arms; it would be more likely to do so by means of ideological persuasion. Finally, it is conceivable that the leading great powers—now the United States and the Soviet Union—might form a kind of world government in partnership, disarm the other powers, and impose peace. But neither of these latter two models would ensure that the dominant powers would not use their strength to exploit the defenseless.

A world government on a federal basis seems to offer the best solution at the present time. But what of the long-term prospects for a humanity in which warlike competition has ceased? Warlike competition has undoubtedly promoted the biological and cultural evolution of human groups. Might not the abolition of this competition lead to stagnation or decay? There are divergent views on this. With the huge pool of genes available at the present time, there should be no immediate danger of genetic decay. Should decay appear, eugenic measures might be a better counter to it than war. The problem of the future genetic development of mankind has been discussed at length by Dobzhansky.[34]

A number of models suggest that change in the internal organization of states might lead to a pacification of the world. It has been argued that the democratic system would lead to peace, for if the decision about peace and war were in the hands of the people, who would be the chief sufferers from war, they would hardly be likely to opt for it. (A discussion of the historical development of this idea is to be found in Fetscher.[35]) But history teaches us that democracies are no more peaceful than other forms of government. The "peace through democracy" theory has hitherto foundered because of the fact that formation of rational opinion by the majority of the population is not possible in prevailing conditions. Public opinion is formed by interest groups (politicians, arms manufacturers, the military) that deceive the electorate by giving them false or one-sided information or, alternatively, do not consult the people but bring about decisions through secret diplomacy. For this reason, socialists seek to establish a homogeneous society. After the elimination of class contradictions, the population would, they claim, be able to make rational decisions about arming or disarming, peace or war, without interest groups being able to bias the formation of opinion. It has, however, been shown in practice that in socialist societies, bureaucrats, party elites, and military—corresponding to the interest groups in democracies—hamper free discussion and thereby the free formation of opinion. Besides, national interests prevail equally in democratic and in socialist states.

The Chinese attempt to prevent the formation of elites by an endless succession of "cultural revolutions" shows that a really egalitarian society can be attained only by force. This suggests that it might be less repressive to tolerate aspirations to rank and property and, by appropriate education and control, to eliminate the harmful effects of the power over other men associated with those things, for both perform ordering functions in social life and stimulate cultural evolution.

The oldest worldwide peace movements have come from the higher religions. The Judeo-Christian doctrine according to which all men are the children of God and equal in his eyes has made a vital contribution to the pacification of the world in the last two thousand years; and together with the other higher religions, it continues to work in the same direction. The religions can claim to their credit that they work for peace by peaceful means. Secular peace movements are still far from this.

Finally, some models are concerned with attaining peace by eliminating individual aggression. As I have mentioned, Lorenz in *On Aggression* takes the view that aggressivity is something we have to live with, since it is a constitutional characteristic of man that at best could be eliminated by the highly problematical method of planned breeding. But aggression could be made innocuous by reorientation. One of the functions ascribed to competitive games is that they fill an important aggression-diverting and binding role. Sipes, however, takes the view that they encourage aggression and should be abolished.[36] As I have pointed out, he fails to distinguish between the short- and the long-term effects of aggression. Competitive games can in fact divert aggression, but at the same time, they train the aggressive system.

Lorenz also shares Freud's view that aggressions can be sublimated and acted out in creative activity. Marcuse sees the root of man's basic aggressive attitude in the domination of the proficiency principle. If this were dropped in favor of a repression-free culture, it would, in his opinion, contribute to the pacification of mankind.[37] According to Wilhelm Reich, an aggression-free society can be attained only by a complete liberation of sexual drives; this has actually led some misguided advocates of frustration-free educational models into encouraging premature childhood sexuality. Meves has rightly pointed out that all too unconsidered experimentation in this field is irresponsible, since the possible ultimate consequences are unknown.[38] Advocates of so-called repression-free education generally overlook the fact that man is by nature a cultural being. The drive life of animals is guided by rigid preexisting controls, but man depends essentially on such controls being transmitted to him culturally. The child expects to be presented with such guidelines, and in their absence, he becomes insecure and—as we now know—really aggressive, for he is left with aggression as his only means of social exploration. He wants to find out where the boundaries lie and by what patterns of order he can orient himself. Children at an early age subject themselves at play to rules they invent for themselves; in other words, they show an obvious appetence for culturation. The attacks on the family that are continually associated with advocacy of a repression-free education are also highly dubious. As I have said elsewhere, it is in the family that the capacity of love of one's neighbor is developed. Children unable to develop a tie with those close to them later have great difficulty in establishing human contacts.

As the advocates of repression-free education pay no regard to man's preexisting dispositions, they run the risk of exposing people to real frustrations. In the family, man certainly develops the ability to distinguish between members and strangers and, with basic trust, also develops basic mistrust. But in the course of evolution, he has also developed the capacity to expand the group at will and to extend the family ethos to those unknown to him.

Some therapeutic proposals aim at damping down individual aggression in general. Training, drugs, surgery, and even selective eugenics have been proposed. I regard all these things as exceedingly dubious while we are still a long way from being able to say whether the abolition of aggression is in fact desirable. The previously mentioned experiments by Dann, who by a concentration test evaluated the "efficiency" of persons whom he angered, are basically irrelevant to the question whether aggression is also correlated with positive qualities. There is a great deal in favor of the assumption that aggressive personalities are well able to put their dynamism to work in a positive way in tackling problems. The therapeutic proposals that aim at reducing the individual's aggressive potential are based on the totally false assumption that the potentially aggressive have a constitutional tendency to destructive aggression. I know many men who "fight" passionately, but not destructively, for peace. Their weapons are arguments.

Without man's attacking spirit, there would certainly be no worthwhile advances in either the intellectual or the social field, and I agree with Hassenstein, who attributes a very high place on the scale of values to civil courage, the courage that enables a defenseless individual to stand up against a powerful enemy.[39] I have also mentioned the positive function of aggression in exploratory contact with the environment.

Therapists' proposals to use the classical conditioning technique, for instance—that is, rewarding desired behavior patterns and punishing undesired (aggressive) behavior by electric shocks or temporary isolation from the group—or other interventionary measures to check individual aggressivity are suitable only in cases of pathological aggressivity (acts of violence). But our problem is not that of the gravely disturbed. What we are concerned with is ridding the world of the phenomenon of perfectly normal young men, who in everyday life are perfectly friendly, going to war and killing their fellows.

This can be done only by an education that does not just damp down aggressions but so socializes them that they are not used destructively. The elimination of barriers to communication and the consequent disappearance of the enemy stereotype can help in this. Also, the child must itself be able to collect experiences with aggression in order to discover its effects on himself and others. In 1970, I pointed out that Freud considered it a crime that educationists did not prepare people for the aggression with which they would later have to cope. And I added that all belittlement of aggression—dismissing it as an acquired habit that could therefore easily be eliminated, for instance—was irresponsible. I was attacked for this by Wintsch,[40] but I maintain that view, which cannot be regarded as advocacy of aggression. In view of man's aggressive potential, socialization can take place only by means of experience with aggression. I am not alone in that opinion. Mead, among others, wrote: "On the world scene, there is increasing evidence that there may be a negative correlation between the amount of experience in childhood with aggressive behavior that falls short of serious damage and the amount of violence that may erupt in that society. Those societies where children receive no training in limited conflict with others have the least experience in halting destruction and killing once it starts."[41]

Here is an example of what Mead referred to: The Semai are noted for their lack of aggressivity. They do not punish their children and abhor violence. But when they suffered losses at the hands of Communists in Malaya and a Semai unit went into action against them, they actually grew drunk with blood. Dentan described what happened:

> Many people who knew the Semai insisted that such an unwarlike people could never make good soldiers. Interestingly enough, they were wrong. Communist terrorists had killed the kinsmen of some of the Semai counterinsurgency troops. Taken out of their nonviolent society and ordered to kill, they seem to have been swept up in a sort of insanity which they call "blood drunkenness." A typical veteran's story runs like this. "We killed, killed, killed. The Malays would stop and go through people's pockets and take their watches and money, we did not think of watches or money. We thought only of killing. Wah, truly we were drunk with blood." One man even told how he had drunk the blood of a man he had killed.[42]

The capacity for self-control is certainly an important condition for peaceful coexistence, and it can be greatly promoted by education. But internal pacification does not automatically mean external pacification, and education must work toward the latter by encouraging a consciousness of humanity and developing a basic attitude of tolerance. Man must learn not automatically to associate appreciation of his own culture with depreciation of other cultures and systems of value, but to recognize and value the right to be different. This can be done by breaking down barriers to communication and ceasing to demonize minorities and alien peoples in the media. Films still present us with models of violence. One's own ("just") wars are glorified, and one's enemies are dehumanized. In this way, enemy dummies are constructed and war is propagated. Such behavior should be sharply condemned.

It is also important to encourage the available counterforces to aggression. In the conciliatory appeals and in the innate behavior patterns of group formation, we have at our disposal very effective controls on aggression that actually enable us to live with it. The great importance of cultural rituals that are ultimately based on them—the feasts, forms of greeting, gift rituals, and finally, good manners—is unfortunately far from being sufficiently recognized. Those who regard the exchange of Christmas presents as merely a product of the consumer society take a very superficial view. Certainly there are such products, but it would be foolish to throw overboard an important binding ritual for that reason.

Peace is an attainable human aim. It corresponds to our inclinations; we seek peace and try to ensure it. It is the goal of the great religions and ideologies, all of which claim to represent the universal interest in this respect; and if their teachings differ, it is not because of any intention to deceive, but because they see reality from different aspects and not always without distortion. But a common aspiration creates a bond over and above ideological frontiers, and that could lead from confrontation to cooperation toward improvement of the international organizations that are already concerned with world peace, just distribution of wealth, protection of the weak, and provision of aid in catastrophes, etc. They must also serve as the ultimate authority in disputes and have jurisdiction to administer their decisions. The right to self-help that has hitherto existed, apart from the right to self-defense, must be surrendered to that authority. The fact that there are still wars is due to, among other reasons, the fact that no such supreme judicial authority yet exists.

War and peace are alternatives for which we are constitutionally equipped. The fact that we see these interconnections gives us the opportunity of making up our mind, and, whichever course we set, the full responsibility is ours.

Conclusion

In many animal species, intraspecific aggression is so ritualized that it does not result in physical harm. This is basically true also of human intragroup aggression, which to a significant extent is based on phylogenetic adaptations and is effectively controlled by such adaptations. Among these are innate signals of submission that prevent aggression between members of a group from escalating into destructiveness. To some extent, a biological norm filter lays down the commandment: "Thou shalt not kill."

Man's intergroup aggression, however, generally aims at destruction. This is the result of cultural pseudospeciation, in the course of which human groups mark themselves off from others by speech and usages, and describe themselves as human and others as not fully human. Underlying this is the innate disposition that manifests itself in the infant's fear and rejection of strangers as well as in primate territoriality, but it is only through man's suppression of his similarly innate inclination to band formation that a sharp demarcation of groups takes place. Thus war developed as a cultural mechanism in the competition of groups (pseudospecies) for space and raw materials. The theory that war first developed in the Neolithic age with the development of horticulture and agriculture does not stand up to critical examination; there is evidence of armed clashes as early as the Paleolithic period. Moreover, present-day hunters and gatherers have been shown to be mostly territorial; they fight for the possession of hunting and gathering grounds.

War is primarily destructive, but as in the development from destructive to ritualized fighting, we see the beginnings of cultural ritualizations of war in the form of conventions that prevent excessive bloodshed. This is obviously of selectionistic advantage. But development in this direction assumes that the functions of war, in the competition for land, for instance, can be performed in some other way, even including war without bloodshed. Among other things, the defeated must be able to retreat, which is a condition for their being spared in the dueling of animals. This is no longer possible for man because there is a shortage of empty spaces, and thus a limit is set to an automatic further humanization of war. But there are other preadaptations that could bring about a cultural development to peace. In the course of cultural pseudospeciation, man has superimposed a cultural norm filter that commands him to kill upon his biological norm filter, which forbids him to kill.

This leads to a conflict of norms, of which man is aware through the conscience that pricks him as soon as he apprehends the enemy that confronts him as a fellow human being. Finally, he shows the same signals that normally have a conciliatory effect in intragroup relations and release sympathy. There is an abundance of observations that reinforce this assumption; victorious warriors, for example, often have to perform penitential rituals before being fully reintegrated into their community.

The root of the universal desire for peace lies in this conflict between cultural and biological norms, which makes men want to bring their biological and cultural norm filters into accord. Our conscience remains our hope, and based on this, a rationally guided evolution could lead to peace. This presupposes recognition of the fact that war performs functions that will have to be performed in some other way, without bloodshed. Those who do not see this and dismiss war as a pathological phenomenon are guilty of a dangerous simplification, since it naturally will not occur to them that those who want peace will have to find a different way of carrying out the functions of war. Man's motivational structure makes him perfectly capable of peaceful coexistence in modern mass society, and education for peace must be primarily education in tolerance in the sense of willingness to understand.

Notes

CHAPTER 1 IS AGGRESSION INEVITABLE?

1. W. Hollitscher, ed., *Aggressionstrieb und Krieg* (Stuttgart, 1973), p. 115.
2. E. Fromm, "Lieber fliehen als kämpfen," *Bild der Wissenschaft* 10 (1974): 52–58.
3. K. Lorenz, *On Aggression* (New York, London, 1966), pp. 29–30. Translation of (1963) *Das sogenannte Böse*.
4. K. Lorenz, "Der Mensch, biologisch gesehen. Eine Antwort an Wolfgang Schmidbauer," *Studium Generale* 24 (1971): 509.
5. I. Eibl-Eibesfeldt, *Der vorprogrammierte Mensch. Das Ererbte als bestimmender Faktor im menschlichen Verhalten* (Vienna, 1973).
6. W. Lepenies, *Frankfurter Allgemeine Zeitung*, April 2, 1974.

CHAPTER 2 SEEKING THE CAUSES OF BEHAVIOR

1. I. Eibl-Eibesfeldt and H. Hass, "Zum Projekt einer ethologisch orientierten Untersuchung menschlichen Verhaltens," *Mitteilungen der Max-Planck-Gesellschaft* 6 (1966): 383–96.
2. Originally published in *Journal für Ornithologie* 83 (1935): 137–413. Translated (1970) in K. Lorenz, *Studies in Animal and Human Behavior*, vol. 1 (Cambridge, Mass.), pp. 101–258.
3. R. A. Hinde, *Animal Behavior: A Synthesis of Ethology and Comparative Psychology* (New York, London, 1966); D. S. Lehrman, "The Presence of the Mate and of Nesting Material as Stimuli for the Development of Incubation Behavior and for Gonadotropic Secretion in the Ring Dove," *Endocrinology* 68 (1961): 507–16.
4. V. G. Dethier and D. Bodenstein, "Hunger in the Butterfly," *Zeitschrift für Tierpsychologie* 15 (1958): 129–40.
5. E. von Holst, "Die relative Koordination als Phänomen und als Methode zentralnervöser Funktionsanalyse," *Ergebnisse der Physiologie* 42 (1939): 228–306. Reprinted (1969) in idem, *Zur Verhaltensphy-*

siologie bei Tieren und Menschen. Gesammelte Abhandlungen, vol. 1 (Munich).

6. Literature in I. Eibl-Eibesfeldt, *Ethology: The Biology of Behavior* (New York, London, 1975). Translation of (1974) *Grundriss der vergleichenden Verhaltensforschung*, 4th ed.

7. H. A. Euler, "Der Effekt von aggressionsabhängiger Strafreizung (Elektroschock) auf das Kampfverhalten von Leghorn Hähnen" (Paper read at 28. Kongress der Deutschen Gesellschaft für Psychologie, Oktober 1972). *Gruppendynamik* 3: 311–18.

8. Lorenz, "Der Kumpan in der Umwelt des Vogels"; E. H. Hess, *Imprinting: Early Experience and the Developmental Psychology of Attachment* (New York, London, 1973); K. Immelmann, "Zur ökologischen Bedeutung prägungsbedingter Isolationsmechanismen," in *Verhandlungen der Deutschen Zoologischen Gessellschaft, 64. Jahresversammlung 1970*, ed. W. Rathmayer (Stuttgart, 1970), pp. 304–14.

9. P. H. Klopfer, "Mother Love: What Turns It On?" *American Scientist* 59 (1971): 404–407.

10. More on this question, which can be referred to only briefly here, can be found in my *Ethology: The Biology of Behavior.*

11. R. W. Sperry, "How a Developing Brain Gets Itself Properly Wired for Adaptive Function," in *The Biopsychology of Development*, ed. E. Tobach, L. R. Aronson, and E. Shaw (New York, 1971), pp. 27–44.

12. For summaries, see K. Lorenz, "Innate Basis of Learning," in *On the Biology of Learning*, ed. H. Pribram (New York, 1969), pp. 13–93, and Eibl-Eibesfeldt, *Ethology: The Biology of Behavior.*

13. W. Ball and F. Tronick, "Infant Responses to Impending Collision: Optical and Real," *Science* 151 (1966): 818–20. See also T. G. Bower, "Slant Perception and Shape Constance in Infants," *Science* 171 (1971): 832–34; and idem, "The Object in the World of the Infant," *Scientific American* 225 (1971): 30–38.

14. B. Hassenstein, "Wesensverschiedene Formen menschlicher Aggressivität in der Sicht der Verhaltensforschung," *Universitas* 28 (1973): 287–95.

15. K. Lorenz, *Behind the Mirror: A Search for a Natural History of Knowledge* (1977). Translation of (1973) *Die Rückseite des Spiegels. Versuch einer Naturgeschichte menschlichen Erkennens.*

16. K. Horn, "Die humanwissenschaftliche Relevanz der Ethologie," in *Kritik der Verhaltensforschung*, ed. G. Roth (Munich, 1974), pp. 190–221. The reference is to R. Spitz and W. G. Cobliner, *The First Year of Life: A Psychoanalytic Study of Normal and Deviant Development of Object Relations* (New York, 1965).

17. K. Lorenz, "Die Entwicklung der vergleichenden Verhaltensforschung in den letzten 12 Jahren," *Zoologischer Anzeiger Supplement* 16 (1953): 36–58.

18. See I. Eibl-Eibesfeldt, *Der vorprogrammierte Mensch. Das Ererbte als bestimmender Faktor im menschlichen Verhalten* (Vienna, 1973).

19. A. Schmidt-Mummendey and H. D. Schmidt, *Aggressives Verhalten* (Munich, 1971), p. 13.

20. W. Hollitscher, ed., *Aggressionstrieb und Krieg* (Stuttgart, 1973), p. 13.

21. W. Schmidbauer, *Biologie und Ideologie: Kritik der Humanethologie* (Hamburg, 1973), p. 43.

22. For information on a number of auxiliary criteria, see W. Wickler, "Über den taxonomischen Wert homologer Verhaltensmerkmale," *Die Naturwissenschaften* 52 (1965): 441–44; idem, "Vergleichende Verhaltensforschung und Phylogenetik," in *Die Evolution der Organismen*, ed. G. Heberer, 3d ed. (Stuttgart, 1967), 1: 420–508; and idem, *Verhalten und Umwelt* (Hamburg, 1972). See also my own works mentioned at the beginning of the book.

23. H. Hass, *Das Energon* (Vienna, 1970).

24. Lorenz, *Behind the Mirror*.

25. E. H. Erikson, "Ontogeny of Ritualization in Man," *Philosophical Transactions of the Royal Society of London*, ser. B, 251 (1966): 337–49.

26. Lorenz, *Behind the Mirror*.

27. O. Koenig, *Kultur und Verhaltensforschung* (Munich, 1970).

28. Hass, *Das Energon*.

29. Eibl-Eibesfeldt, *Der vorprogrammierte Mensch*; and "Medlpa (Mbowamb)–Neuguinea–Werberitual (Amb Kanànt)," *Homo* 25 (1974): 274–84.

30. Wickler, *Verhalten und Umwelt*.

31. H. Sbrzesny, "Die Spiele der !Ko-Buschleute. Unter besonderer Berücksichtigung ihrer sozialisierenden und gruppenbindenden Funktion" (Thesis, Faculty of Biology, Munich University, 1975).

CHAPTER 3 AGGRESSION WITHIN THE GROUP

1. H. D. Dann, *Aggression und Leistung* (Stuttgart, 1972).

2. A. H. Buss, *The Psychology of Aggression* (New York, 1961), p. 1.

3. H. Selg, *Diagnostik der Aggressivität* (Göttingen, 1968), p. 22.

4. J. Dollard, L. W. Doob, N. E. Miller, O. H. Mowrer, and R. R. Sears, *Frustration and Aggression* (New Haven, 1939), p. 9.

5. L. Berkowitz, *Aggression: A Social-Psychological Analysis* (New York, London, 1962). Other authors who include intention in their definition are S. Feshbach, "The Function of Aggression and the Regulation of Aggressive Drive," *Psychological Review* 71 (1964): 257–72; and F. Merz, "Aggression und Aggressionstrieb," in *Handbuch der Psychologie*, ed. H. Thomae, vol. 2, *Motivation* (Göttingen, 1965), pp. 569–601.

6. C. F. Graumann, *Einführung in die Psychologie* (Frankfurt, 1969), p. 85.

7. H. Markl, "Die Evolution des sozialen Lebens der Tiere," in *Verhaltensforschung*, ed. K. Immelmann (supplementary volume to Bernhard Grzimek, *Tierleben* [Munich, 1974]), p. 475.

8. L. H. Phillips and M. Konishi, "Control of Aggression by Singing in Crickets," *Nature* 241 (1972): 64–65.

9. H. Kummer, "Aggression bei Affen," in *Der Mythos vom Aggressionstrieb*, ed. A. Plack (Munich, 1973), p. 71.

10. R. N. Johnson, *Aggression in Man and Animals* (Philadelphia, London, 1972), p. 6.

11. K. E. Moyer, "Experimentale Grundlagen eines physiologischen Modells agressiven Verhaltens," in *Aggressives Verhalten,* ed. A. Schmidt-Mummendey and H. D. Schmidt (Munich, 1971) and *The Physiology of Hostility* (Chicago, 1971).

12. D. J. Reis, "Central Neurotransmitters in Aggression," in *Aggression,* ed. S. H. Frazier, *Proceedings of the Association, Research Publications of the Association for Nervous and Mental Disease* (Baltimore, 1974), pp. 119–48.

13. See P. Leyhausen, *Verhaltensstudien an Katzen,* 3d ed. (Berlin, 1973).

14. R. Ardrey, *The Territorial Imperative* (New York, 1966); R. A. Dart, "The Bone-bludgeon Hunting Technique of Australopithecus," *South African Science* 2 (1949): 150–52; and idem, "The Predatory Transition from Ape to Man," *International Anthropological and Linguistic Review* 1 (1953): 201–18.

15. J. P. Scott, *Aggression* (Chicago, 1958).

16. E. von Holst and U. von Saint-Paul, "Vom Wirkungsgefüge der Triebe," *Die Naturwissenschaften* 18 (1960): 409–22.

17. R. W. Hunsperger, "Reizversuche im periventrikulären Grau des Mittel- und Zwischenhirns" (film), *Helvetica Physiologica Acta* 12 (1954): C4–C6.

18. F. Hacker, *Aggression* (Vienna, 1971), p. 79.

19. Dann, *Aggression und Leistung.*

20. I. Eibl-Eibesfeldt, "Zur Ethologie des Hamsters (*Cricetus cricetus L.*)," *Zeitschrift für Tierpsychologie* 10 (1953): 204–54.

21. G. B. Schaller, *The Serengeti Lion: A Study of Predator-Prey Relations* (Chicago, 1972).

22. T. Schultze-Westrum, *Biologie des Friedens: Auf dem Weg zu solidarischem Verhalten* (Munich, 1974).

23. I. Eibl-Eibesfeldt, *Land of a Thousand Atolls* (New York, London, 1967). Translation of (1964) *Im Reich der tausend Atolle.*

24. K. Lorenz, "Die angeborenen Formen möglicher Erfahrung," *Zeitschrift für Tierpsychologie* 5 (1943): 235–409.

25. H. Markl, "Aggression und Beuteverhalten bei Piranhas (*Serrasalminae*)," *Zeitschrift für Tierpsychologie* 30 (1972): 190–216.

26. F. R. Walther, "Entwicklungszüge im Kampf- und Parrungsverhalten der Horntiere," *Jahrbuch Georg von Opel Freigehege für Tierforschung* 3 (1961): 90–115; and idem, *Mit Horn und Huf* (Berlin, 1966).

27. J. M. Deag and J. H. Crook, "Social Behaviour and 'Agonistic Buffering' in the Wild Barbary Macaque *Macaca sylvana,*" *Folia Primatologica* 15 (1971): 183–200.

28. F. A. Pitelka, "Numbers, Breeding Schedule and Territory in Pectoral Sandpipers of Northern Alaska," *Condor* 61 (1959): 233–64.

29. E. O. Wilson, "Competitive and Aggressive Behavior," in *Man*

and Beast: Comparative Social Behavior, Smithsonian Annual 3, ed. J. F. Eisenberg and W. S. Dillon (Washington, D.C., 1971), p. 195.

30. E. O. Willis, *The Behavior of the Bicolored Antbird*, University of California Publications in Zoology no. 79 (Berkeley, 1967).

31. H. Kummer, "Spacing Mechanisms in Social Behavior," in *Man and Beast: Comparative Social Behavior*, Smithsonian Annual 3, ed. J. F. Eisenberg and W. S. Dillon (Washington, D.C., 1971), p. 223.

32. W. D. Hamilton, "The Genetical Evolution of Social Behavior," *Journal of Theoretical Biology* 7 (1964): 1–52.

33. T. Schjelderup-Ebbe, "Beiträge zur Sozialpsychologie des Haushuhns," *Zeitschrift für Psychologie* 88 (1922): 225–52; and idem, "Soziale Verhältnisse bei Vögeln," *Zeitschrift für Psychologie* 90 (1922): 106–107.

34. M. R. A. Chance, "Attention Structure as the Basis of Primate Rank Orders," *Man* 2 (1967): 503–18.

35. B. Hold, "Rangordnungsverhalten bei Vorschulkindern. Eine vorläufige Mitteilung," *Homo* 25 (1974): 252–67.

36. I. Eibl-Eibesfeldt, "Angeborenes und Erworbenes im Verhalten einiger Säuger," *Zeitschrift für Tierpsychologie* 20 (1963): 705–54.

37. D. Freeman, "Aggression: Instinct or Symptom?" *Australian & New Zealand Journal of Psychiatry* 5 (1971): 71.

38. Literature in I. Eibl-Eibesfeldt, *Ethology: The Biology of Behavior* (New York, London, 1975). Translation of (1974) *Grundriss der vergleichenden Verhaltensforschung*, 4th ed.

39. D. Lack, *The Life of the Robin* (Cambridge, 1943).

40. J. H. Mackintosh and E. C. Grant, "The Effect of Olfactory Stimuli on the Agonistic Behaviour of Laboratory Mice," *Zeitschrift für Tierpsychologie* 23 (1966): 584–87.

41. C. Y. Leong, "The Quantitative Effect of Releasers on the Attack Readiness of the Fish *Haplochromis burtoni* (Cichlidae, Pisces)," *Zeitschrift für vergleichende Physiologie* 65 (1969): 29–50.

42. T. I. Thompson, "Visual Reinforcement in Siamese Fighting Fish," *Science* 141 (1963): 55–57; and idem, "Visual Reinforcement in Fighting Cocks," *Journal of Experimental Analysis of Behavior* 7 (1964): 45–49.

43. A. Tellegen, J. M. Horn, and R. G. Legrand, "Opportunity for Aggression as Reinforcer in Mice," *Psychonomic Science* 14 (1969): 104–105.

44. Von Holst and von Saint-Paul, "Vom Wirkungsgefüge der Triebe."

45. J. M. R. Delgado, "Evoking and Inhibiting Aggressive Behavior by Radiostimulation in Monkey Colonies," *American Zoologist* 6 (1966): 669–81; and idem, "Radiostimulation of the Brain in Primates and in Man," *Anesthesia and Analgesia Current Researches* 48 (1969): 529–43.

46. R. Plotnik, "Brain Stimulation and Aggression: Monkeys, Apes, and Humans," in *Primate Aggression, Territoriality, and Xenophobia: A Comparative Perspective*, ed. R. L. Holloway (New York, London, 1974), pp. 389–415.

47. J. Kruijt, "Ontogeny of Social Behaviour in Burmese Red Jungle Fowl (*Gallus gallus spadiceus*)," *Behaviour*, Supplement 12.

48. Moyer, "Experimentale Grundlagen eines physiologischen Modells aggressiven Verhaltens," p. 50.

49. O. A. E. Rasa, "The Effect of Pair Isolation on Reproductive Success in *Etroplus maculatus* (Cichlidae)," *Zeitschrift für Tierpsychologie* 26 (1969): 846–52.

50. W. Wickler, "Soziales Verhalten als ökologische Anpassung," in *Verhandlungen der Deutschen Zoologischen Gesellschaft, 64. Jahresversammlung 1970*, ed. W. Rathmayer (Stuttgart, 1970), pp. 291–304.

51. O. A. E. Rasa, "Appetence for Aggression in Juvenile Damsel Fish," *Zeitschrift für Tierpsychologie*, Supplement 7.

52. Wickler, "Soziales Verhalten als ökologische Anpassung."

53. I. Goldenbogen, "Uber den Einfluss sozialer Isolation auf die aggressive Handlungsbereitschaft von *Xiphophorus helleri* (Poeciliidae) und *Haplochromis burtoni* (Cichlidae)," *Zeitschrift für Tierpsychologie* 44 (1974): 25–44.

54. D. Franck and U. Wilhelmi, "Veränderungen der aggressiven Handlungsbereitschaft männlicher Schwertträger, *Xiphophorus helleri*, nach sozialer Isolation," *Experientia* 29 (1973): 896–97.

55. E. Courchesne and G. W. Barlow, "Effect of Isolation on Components of Aggressive and Other Behavior in the Hermit Crab *Pagurus samuelis*," *Zeitschrift für Vergleichende Physiologie* 75 (1971): 32–48.

56. L. Valzelli, "Aggressive Behaviour Induced by Isolation," in *Aggressive Behaviour*, ed. S. Garattini and E. B. Sigg (Amsterdam, 1969), pp. 70–76; and K. Lagerspetz, "Interrelations in Aggression Research: A Synthetic Overview," *Psykologiska Rapporter* 4 (Åbo, Finland).

57. R. B. Cairns, "Fighting and Punishment from a Developmental Perspective," in *Nebraska Symposium on Motivation*, ed. J. K. Cole and D. D. Jensen (Lincoln, 1972), pp. 59–124.

58. W. Heiligenberg, "Ein Versuch zur ganzheitsbezogenen Analyse des Instinktverhaltens eines Fisches (*Pelmatochromis subocellatus kribensis, Boul., Cichlidae*)," *Zeitschrift für Tierpsychologie* 21 (1964): 1–52: idem and U. Kramer, "Aggressiveness as a Function of External Stimulation," *Journal of Comparative Physiology* 77 (1972): 332–40.

59. Wickler, "Soziales Verhalten als ökologische Anpassung," pp. 294ff.

60. K. D. Roeder, "Spontaneous Activity and Behavior," *Scientific Monthly* 80 (1955): 362–70; T. H. Bullock and G. A. Horridge, *Structure and Function in the Nervous System of Invertebrates*, vols. 1 and 2 (San Francisco, 1965).

61. M. Jouvet, "Le Discours Biologique," *La Revue de Médicine* 16–17 (1972): 1003–63.

62. Lagerspetz, "Interrelations in Aggression Research."

63. K. Lagerspetz, "Aggression and Aggressiveness in Laboratory Mice," in *Aggressive Behaviour*, ed. S. Garattini and E. B. Sigg (Amsterdam, 1969), pp. 77–85.

CHAPTER 4 THE "PEACEFUL" PRIMATE: TERRITORY AND AGGRESSION

1. V. Reynolds, "Open Groups in Human Evolution," *Man* 1 (1966): 441–52; V. Reynolds and F. Reynolds, "Chimpanzees of the Budongo Forest," in *Primate Behavior: Field Studies of Monkeys and Apes*, ed. I. DeVore (New York, 1965); J. Goodall, "Chimpanzees of the Gombe Stream Reserve," in ibid.

2. J. van Lawick-Goodall, "The Behaviour of the Chimpanzee," in *Hominisation und Verhalten*, ed. G. Kurth and I. Eibl-Eibesfeldt (Stuttgart, 1975), p. 81.

3. K. Izawa, "Unit Groups of Chimpanzees and Their Nomadism in the Savanna Woodland," *Primates* 11 (1970): 1–46; T. Nishida, "The Social Group of Wild Chimpanzees of the Mahali Mountains," *Primates* 9 (1968): 167–224; Y. Sugiyama, "Social Behavior of Chimpanzees in the Budongo Forest, Uganda," *Primates* 10 (1969): 197–225; J. Itani and A. Suzuki, "The Social Unit of Chimpanzees," *Primates* 8 (1967): 355–81.

4. Reynolds and Reynolds, "Chimpanzees of the Budongo Forest," p. 396.

5. Itani and Suzuki, "The Social Unit of Chimpanzees."

6. T. Nishida and K. Kawanaka, "Inter-Unit Group Relationships among Wild Chimpanzees of the Mahali Mountains," *Kyoto University African Studies* no. 7 (1972), pp. 131–69.

7. Lawick-Goodall, "The Behaviour of the Chimpanzee."

8. J. D. Bygott, "Cannibalism among Wild Chimpanzees," *Nature* 238 (1972): 410–11.

9. J. van Lawick-Goodall, *In the Shadow of Man* (Boston, London, 1971), p. 125.

10. Nishida and Kawanaka made similar observations: see their "Inter-Unit Group Relationships among Wild Chimpanzees of the Mahali Mountains."

11. Lawick-Goodall, "The Behaviour of the Chimpanzee."

12. Lawick-Goodall, *In the Shadow of Man*, p. 222.

13. A. Kortlandt, "Chimpanzees in the Wild," *Scientific American* 206 (5) (1962): 128–38; "Bipedal Armed Fighting in Chimpanzees," *Proceedings of 16th International Congress of Zoologists*, vol. 3 (1963): 64; "Handgebrauch bei freilebenden Schimpansen," in *Handgebrauch und Verständigung bei Affen und Frühmenschen*, ed. B. Rensch (Berne, 1967), pp. 59–102. See further discussion of Kortlandt's findings on p. 74.

14. H. Albrecht and S. C. Dunnett, *Chimpanzees in Western Africa* (Munich, 1971).

15. V. Reynolds and O. Luscombe, "Chimpanzee Rank Orders and the Function of Displays," *Proceedings of 2d International Congress on Primatology, Zurich 1968* (Basel, 1969).

16. J. A. R. A. M. van Hooff, *Aspecten van het Sociale Gedrag en de Communicatie bij Humane en Hogere Niet-Humane Primaten (Aspects of Social Behavior and Communication in Human and Higher Non-Human Primates)* (Rotterdam, 1971).

17. G. B. Schaller, *The Mountain Gorilla: Ecology and Behavior* (Chicago, 1963); D. Fossey, "More Years with Mountain Gorillas," *National Geographic Magazine* 140 (1971): 574–85; M. Kawai and H. Mizuhara, "An Ecological Study of the Wild Mountain Gorilla," *Primates* 2 (1959): 1–42.

18. C. R. Carpenter, *A Field Study in Siam of the Behavior and Social Relations of the Gibbon*, Comparative Psychology Monographs no. 16 (1940). Reprinted (1964) in idem, *Naturalistic Behavior of Nonhuman Primates* (University Park, Pa.), pp. 145–271.

19. W. Schmidbauer, ed., *Evolutionstheorie und Verhaltensforschung* (Hamburg, 1974), p. 178. The reference is to Reynolds and Reynolds, "Chimpanzees of the Budongo Forest."

20. Ibid., p. 179.

21. G. Teleki, *The Predatory Behavior of Wild Chimpanzees* (Lewisburg, Pa., 1973).

22. Bygott, "Cannibalism among Wild Chimpanzees," p. 411.

23. A. Kortlandt, *New Perspectives on Ape and Human Evolution* (Amsterdam, 1972).

24. Teleki, *The Predatory Behavior of Wild Chimpanzees.*

25. Kortlandt, *New Perspectives on Ape and Human Evolution.*

26. A. Kortlandt and M. Kooij, "Protohominid Behavior in Primates," *Symposia of the Zoological Society of London* 10 (1963): 61–68.

27. Kortlandt, *New Perspectives on Ape and Human Evolution*, p. 82.

CHAPTER 5 AGGRESSION IN MAN

1. P. D. Nesbitt and G. Steven, "Personal Space and Stimulus Intensity at a Southern California Amusement Park, *Sociometry* 37 (1974): 105–15.

2. I. Eibl-Eibesfeldt, *Love and Hate: The Natural History of Behavior Patterns* (New York, London, 1974). Translation of (1970) *Liebe und Hass. Zur Naturgeschichte elementarer Verhaltensweisen.*

3. N. J. Felipe and R. Sommer, "Invasions of Personal Space," *Social Problems* 14 (1966): 206–14.

4. A. H. Esser, "Interactional Hierarchy and Power Structure on a Psychiatric Ward," in *Behaviour Studies in Psychiatry*, ed. S. J. Hutt and C. Hutt (Oxford, New York, 1970), pp. 25–59; R. J. Palluck and A. H. Esser, "Controlled Experimental Modification of Aggressive Behavior in Territories of Severely Retarded Boys," *American Journal of Mental Deficiency* 76 (1971): 23–29; and idem, "Territorial Behavior as an Indicator of Changes in Clinical Behavioral Condition of Severely Retarded Boys," *American Journal of Mental Deficiency* 76 (1971): 284–90.

5. J. J. Edney, "Property, Possession and Permanence: A Field Study in Human Territoriality," *Journal of Applied Social Psychology* 2 (1972): 275–82.

6. J. J. Edney and N. L. Jordan-Edney, "Territorial Spacing on a Beach," *Sociometry* 37 (1974): 92–104.

7. H. Kummer, W. Götz, and W. Angst, "Triadic Differentiation: An

Inhibitory Process Protecting Pair Bonds in Baboons," *Behaviour* 49 (1974): 62–67.

8. M. Sherif and C. W. Sherif, *Groups in Harmony and Tension* (New York, 1966).

9. M. R. A. Chance, "Attention Structure as the Basis of Primate Rank Orders," *Man* 2 (1967): 503–18.

10. V. Packard, *The Pyramid Climbers* (New York, 1962).

11. D. Morris, *The Human Zoo* (London, New York, 1969).

12. Morris, ibid.

13. D. J. Stayton, R. Hogan, and M. D. S. Ainsworth, "Infant Obedience and Maternal Behavior: The Origins of Socialization Reconsidered," *Child Development* 42 (1971): 1066.

14. S. Milgram, "Einige Bedingungen von Autoritätagehorsam und seiner Verweigerung," *Zeitschrift für experimentelle und angewandte Psychologie* 13 (1966): 433–63.

15. H. Schoeck, *Der Neid* (Freiburg, Munich, 1966).

16. B. Hold, "Rangordnungsverhalten bei Vorschulkindern, Eine vorläufige Mitteilung," *Homo* 25 (1974): 252–67.

17. B. Hassenstein, *Verhaltensbiologie des Kindes* (Munich, 1973).

18. B. Hassenstein, "Wesensverschiedene Formen menschlicher Aggressivität in der Sicht der Verhaltensforschung," *Universitas* 28 (1973): 291.

19. R. L. Trivers, "The Evolution of Reciprocal Altruism," *Quarterly Review of Biology* 46 (1971): 35–57.

20. P. E. McGhee, "On the Cognitive Origins of Incongruity Humor: Fantasy Assimilation versus Reality Assimilation," in *The Psychology of Humor: Theoretical Perspectives and Empirical Issues*, ed. J. H. Goldstein and P. E. McGhee (New York, London, 1972).

21. K. Lorenz, *On Aggression* (New York, London, 1966). Translation of (1963) *Das sogenannte Böse*; I. Eibl-Eibesfeldt, *Ethology: The Biology of Behavior* (New York, London, 1974). Translation of (1974) *Grundriss der vergleichenden Verhaltensforschung*, 4th ed.

22. M. K. Rothbart, "Laughter in Young Children," *Psychological Bulletin* 80 (1973): 247-56.

23. J. A. R. A. M. van Hooff, *Aspecten van het Sociale Gedrag en de Communicatie bij Humane en Hogere Niet-Humane Primaten (Aspects of Social Behavior and Communication in Human and Higher Non-Human Primates)* (Rotterdam, 1971).

24. H. Basedow, "Anthropological Notes on the Western Coastal Tribes of the Northern Territory of South Australia," *Transactions of the Royal Society of South Australia* 31 (1906): 1–62.

25. I. Eibl-Eibesfeldt, *Der vorprogrammierte Mensch. Das Ererbte als bestimmender Faktor im menschlichen Verhalten* (Vienna, 1973).

26. R. A. Baron, "Sexual Arousal and Physical Aggression: The Inhibiting Effect of 'Cheesecake' and Nudes," *Bulletin of the Psychonomic Society* 3 (1974): 337–39; idem, "The Aggression-Inhibiting Influence of Heightened Sexual Arousal," *Journal of Personality and Social Psychology* 30 (1974): 318–22.

27. See also R. A. Baron and R. L. Ball, "The Aggression-Inhibiting Influence of Non-Hostile Humor," *Journal of Experimental Social Psychology* 10 (1974): 23–33.

28. Eibl-Eibesfeldt, *Der vorprogrammierte Mensch.*

29. K. Horn, "Die humanwissenschaftliche Relevanz der Ethologie," in *Kritik der Verhaltensforschung*, ed. G. Roth (Munich, 1974), p. 207.

30. Ibid.

31. C. Russell and W. M. S. Russell, *Violence, Monkeys and Man* (London, New York, 1969/70).

32. J. E. Hokanson and S. Shetler, "The Effect of Overt Aggression on Physiological Tension Level," *Journal of Abnormal and Social Psychology* 63 (1961): 446–48.

33. S. Milgram, "Behavioral Study of Obedience," *Journal of Abnormal and Social Psychology* 67 (1963): 372–78; and idem, *Obedience to Authority: An Experimental View* (New York, 1974).

34. Stayton, Hogan, and Ainsworth, "Infant Obedience and Maternal Behavior."

35. G. F. Vicedom and H. Tischner, *Die Kultur der Hagenbergstämme*, vol. 1, *Die Mbowamb* (Hamburg, 1943/48).

36. H. König, "Der Rechtsbruch und sein Ausgleich bei den Eskimos," *Anthropos* 20 (1925): 314–15.

37. L. von Hörmann, *Tiroler Volkstypen* (Vienna, 1877).

38. F. Lüers, "Volkskundliches aus Steinberg beim Achensee in Tirol," *Bayerische Hefte für Volkskunde* 6 (1919): 106–30.

39. T. Kochman, "Toward an Ethnography of Black American Speech Behavior," in *Afro-American Anthropology: Contemporary Perspectives*, ed. N. E. Whitten and J. F. Szwed (New York, 1970), p. 159.

40. Eibl-Eibesfeldt, *Love and Hate* and *Der vorprogrammierte Mensch.*

41. Private communication to the author from Dr. Schmidt Dumont, Kabul.

42. M. Shokeid, "Conflict and Entertainment: An Analysis of Social Gatherings and Celebrations Among Moroccan Immigrants in Israel" (Paper read at 9th International Congress of Anthropological and Ethnological Sciences, August–September 1973, Chicago).

43. A. R. Radcliffe-Brown, "On Joking-Relationships," *Africa* 13 (1940): 195–210.

44. S. F. Nadel, *The Foundations of Social Anthropology* (Glencove, Ill., London, 1951).

45. L. von Hörmann, *Tiroler Volkstypen.*

46. P. Weidkuhn, "Aggressivität und Normativität. Uber die Vermittlerrolle der Religion zwischen Herrschaft und Freiheit. Ansätze zu einer kulturanthropologischen Theorie der sozialen Norm," *Anthropologie* 63/64 (1968/69): 361–94; and "Fastnacht–Revolte–Revolution." *Zeitschrift für Religion und Geistesgeschichte* 21 (1969): 289.

47. U. Oberem, "Zur Geschichte des lateinamerikanischen Landarbeiters: Conciertos und Huasipungueros in Ecuador," *Anthropos* 62 (1967): 759–88.

48. M. V. Young, *Fighting with Food–Leadership, Values and Social Control in a Massim Society* (Cambridge, 1971).
49. R. Spitz, "Anxiety in Infancy," *International Journal of Psycho-Analysis* 31 (1950): 139–43.
50. Eibl-Eibesfeldt, *Ethology: The Biology of Behavior.*
51. D. W. Ploog, J. Blitz, and F. Ploog, "Studies on Social and Sexual Behavior of the Squirrel Monkey (*Saimiri sciureus*)," *Folia Primatologica* 1 (1963): 29–66; W. Wickler, "Ursprung und biologische Deutung des Genitalpräsentierens männlicher Primaten," *Zeitschrift für Tierpsychologie* 23 (1966): 422–37.
52. Hokanson and Shetler, "The Effect of Overt Aggression on Physiological Tension Level."
53. S. Feshbach, "The Stimulating versus Cathartic Effects of a Vicarious Aggressive Activity," *Journal of Abnormal and Social Psychology* 63 (1961): 381–85.
54. L. Berkowitz, "Aggressive Humor as a Stimulus to Aggressive Responses," *Journal of Personality and Social Psychology* 16 (1970): 710–17; E. S. Dworkin and J. S. Efran, "The Angered: Their Susceptibility to Varieties of Humor," *Journal of Personality and Social Psychology* 6 (1967): 233–36; D. Singer, "Aggression Arousal, Hostile Humor, Catharsis," *Journal of Personality and Social Psychology* Monograph Supplement 8, vol. 1, pt. 2.
55. D. Landy and D. Mattee, "Evaluation of an Aggressor as a Function of Exposure to Cartoon Humor," *Journal of Personality and Social Psychology* 9 (1969): 237–41.
56. Baron and Ball, "The Aggression-Inhibiting Influence of Non-Hostile Humor."
57. A. Bandura, *Aggression: A Social Learning Analysis* (New York, 1973); Berkowitz, "Aggressive Humor as a Stimulus to Aggressive Responses"; L. Berkowitz, R. Corwin, and M. Heironimus, "Film Violence and Subsequent Aggressive Tendencies," *Public Opinion Quarterly* 27 (1963): 217–29; S. Feshbach and R. Singer, *Television and Aggression* (San Francisco, 1971); R. G. Sipes, "War, Sports and Aggression: An Empirical Test of Two Rival Theories," *American Anthropologist* 75 (1973): 64–86.
58. Most recently in Eibl-Eibesfeldt, *Der vorprogrammierte Mensch.*
59. Sipes, "War, Sports and Aggression: An Empirical Test of Two Rival Theories."
60. H. Sbrzesny, "Die Spiele der !Ko-Buschleute unter besonderer Berücksichtigung ihrer sozialisierenden und gruppenbindenden Funktionen," *Monographien zur Humanethologie* 2 (Munich-Zurich, 1976).
61. Sipes, "War, Sports and Aggression: An Empirical Test of Two Rival Theories," p. 80.
62. W. Michaelis, "Der Aggressions-'Trieb' im Streit der Zoologie und Psychologie," *Naturwissenschaftliche Rundschau* 27 (1974): 253–65; Lorenz, *On Aggression.*
63. L. Berkowitz, *Aggression: A Social-Psychological Analysis* (New York, London, 1962).

64. J. E. Hokanson, "Psychophysiological Evaluation of the Catharsis Hypothesis," in *The Dynamic of Aggression*, ed. E. I. Megargee and J. E. Hokanson (New York, 1970).

65. Hokanson and Shetler, "The Effect of Overt Aggression on Physiological Tension Level."

66. A. H. Buss, *The Psychology of Aggression* (New York, 1961); Feshbach, "The Stimulating versus Cathartic Effects of a Vicarious Aggressive Activity."

67. S. K. Mallick and B. R. McCandless, "A Study of Catharsis of Aggression," *Journal of Personality and Social Psychology* 4 (1966): 591–96.

68. J. W. Baker and K. W. Schaie, "Effects of Aggression 'Alone' or 'with Another' on Physiological and Psychological Arousal," *Journal of Personality and Social Psychology* 12 (1969): 80–96; D. Bramel, B. Taub, and B. Blum, "An Observer's Reaction to the Suffering of His Enemy," *Journal of Personality and Social Psychology* 8 (1968): 384–92.

69. A. N. Doob and L. Wood, "Catharsis and Aggression: The Effects of Annoyance and Retaliation on Aggressive Behavior," *Journal of Personality and Social Psychology* 28 (1972): 156–62; V. J. Konečni, "The Mediation of Aggressive Behavior: Arousal Level Versus Anger and Cognitive Labeling," *Journal of Personality and Social Psychology* 32 (1975): 706–12; V. J. Konečni and A. N. Doob, "Catharsis through Displacement of Aggression," *Journal of Personality and Social Psychology* 23 (1972): 379–87.

70. R. G. Geen, D. Stonner, and G. L. Shope, "The Facilitation of Aggression by Aggression: Evidence against the Catharsis Hypothesis," *Journal of Personality and Social Psychology* 31 (1975): 721–26.

71. V. J. Konečni and E. B. Ebbesen, "Disinhibition vs. the Cathartic Effect: Artifact and Substance," *Journal of Personality and Social Psychology* 34 (1976): 352–65.

72. A. Plack, *Die Gessellschaft und das Böse*, 2d ed. (Munich, 1968).

73. F. A. Gibbs, "Ictal and Non-Ictal Psychiatric Disorders in Temporal Lobe Epilepsy," *Journal of Nervous and Mental Disease* 113 (1951): 522–28; K. E. Moyer, "Internal Impulses to Aggression," *Transactions of the New York Academy of Science*, 2d. ser., 31 (1968/69): 104–14; idem, "Experimentale Grundlagen eines physiologischen Modells aggressiven Verhaltens," in *Aggressives Verhalten*, ed. A. Schmidt-Mummendey and H. D. Schmidt (Munich, 1971); idem, *The Physiology of Hostility* (Chicago, 1971); and W. H. Sweet, F. Ervin, and V. H. Mark, "The Relationship of Violent Behavior to Focal Cerebral Disease," in *Aggressive Behaviour*, ed. S. Garattini and E. B. Sigg (Amsterdam, 1969), pp. 336–52.

74. K. D. Roeder, "Spontaneous Activity and Behavior," *Scientific Monthly* 80 (1955): 362–70; T. H. Bullock and G. A. Horridge, *Structure and Function in the Nervous System of Invertebrates*, vols. 1 and 2 (San Francisco, 1965).

75. A. Bandura and R. H. Walters, *Social Learning and Personality Development* (New York, 1963).

76. D. J. Hicks, "Imitation and Retention of Film-Mediated Aggressive Peer and Adult Models," *Journal of Personality and Social Psychology* 2 (1965): 97–100; R. H. Walters and E. L. Thomas, "Enhancement of Punitiveness by Visual and Audiovisual Displays," *Canadian Journal of Psychology* 17 (1963): 244–55.

77. Summarized in Feshbach and Singer, *Television and Aggression.*

78. N. A. Chagnon, "Yanomamö Social Organization and Warfare," in *War: The Anthropology of Armed Conflict and Aggression*, ed. M. Fried, M. Harris, and R. Murphy (Garden City, N.Y., 1968), pp. 130–31.

79. Sbrzesny, "Die Spiele der !Ko-Buschleute unter besonderer Berücksichtigung ihrer sozialisierenden und gruppenbindenden Funktionen."

80. Hassenstein, *Verhaltensbiologie des Kindes* and see p. 87.

81. Michaelis, "Der Aggressions-'Trieb' im Streit der Zoologie und Psychologie."

CHAPTER 6 INTERGROUP AGGRESSION AND WAR

1. A. Toynbee, "Tradition und Instinkt," in *Vom Sinn der Tradition*, ed. L. Reinisch (Munich, 1966), p. 39.

2. R. S. Bigelow, *The Dawn Warriors: Man's Evolution Toward Peace* (London, 1970).

3. E. Fromm, *The Anatomy of Human Destructiveness* (New York, London, 1973).

4. R. Conradt, "Intergruppenaggression—ein artspezifisches Merkmal des Menschen," *Universitas* 28 (1973):1017.

5. A. Montagu, ed., *Man and Aggression* (New York, Toronto, 1968).

6. V. Reynolds, "Open Groups in Human Evolution," *Man* 1 (1966): 449. The reference is to H. V. Vallois, "The Social Life of Early Man: The Evidence of Skeletons," in *Social Life of Early Man*, ed. S. L. Washburn (Chicago, 1961), 229.

7. E. S. Rogers, "Band Organization among the Indians of Eastern Subarctic Canada," *Bulletin of the National Museum of Canada* no. 288 (1969), pp. 21–55.

8. R. A. Dart, "The Bone-bludgeon Hunting Technique of Australopithecus," *South African Science* 2 (1949): 150–52; idem, "The Predatory Transition from Ape to Man," *International Anthropological and Linguistic Review* 1 (1953): 201–18.

9. M. K. Roper, "A Survey of Evidence for Intrahuman Killing in the Pleistocene," *Current Anthropology* 10 (1969): 427–59.

10. A. Mohr, "Häufigkeit und Lokalisation von Frakturen und Verletzungen am Skelett vor- und frühgeschichtlicher Menschengruppen," *Ethnographisch-Archäologische Zeitschrift* 12 (1971): 139–42.

11. H. Kühn, *Kunst und Kultur der Vorzeit. Das Paläolithikum* (Berlin, 1929); idem, *Auf den Spuren des Eiszeitmenschen* (Munich, 1958).

12. Kühn, *Auf den Spuren des Eiszeitmenschen*, p. 105.

13. H. Behrens, "Historische Bewegkräfte im Neolithikum Mitteleuropas," *Archaeologia Austriaca* 55 (1974): 91–94.

14. R. S. Bigelow, "The Evolution of Cooperation, Aggression and Self-Control," in *Nebraska Symposium on Motivation*, ed. J. K. Cole and D. D. Jensen (Lincoln, 1972), p. 8.

15. H. Helmuth, "Zum Verhalten des Menschen: Die Aggression," *Zeitschrift für Ethnologie* 92 (1967): 265–73; R. B. Lee, "What Hunters Do for a Living," in *Man the Hunter*, ed. R. B. Lee and I. DeVore (Chicago, 1968), pp. 30–48; M. D. Sahlins, "The Origin of Society," *Scientific American* 204 (1906): 76–87; J. Woodburn, "Stability and Flexibility in Hadza Residential Groupings," in *Man the Hunter*, pp. 103–10.

16. I. DeVore, "The Evolution of Human Society," in *Man and Beast: Comparative Social Behavior*, Smithsonian Annual 3, ed. J. F. Eisenberg and W. S. Dillon (Washington, 1971), p. 310.

17. M. G. Bicchieri, ed., *Hunters and Gatherers Today: A Socio-economic Study of Eleven Such Cultures in the Twentieth Century* (New York, 1972).

18. E. R. Service, *Primitive Social Organization: An Evolutionary Perspective* (New York, 1962).

19. W. T. Divale, "System Population Control in the Middle and Upper Palaeolithic: Inferences Based on Contemporary Hunter-Gatherers," *World Archaeology* 4 (1972): 222–43.

20. L. T. Hobhouse, "The Simplest Peoples, pt. 2, Peace and Order among the Simplest Peoples," *British Journal of Sociology* 7 (1956): 96–119; L. Frobenius, *Weltgeschichte des Krieges* (Jena, 1903).

21. H. Schjelderup, *Einführung in die Psychologie* (Bern, 1963).

22. F. Boas, "The Central Eskimo," *Sixth Annual Report of the Bureau of American Ethnology, 1884–85* (Washington, D.C., 1888), pp. 399–669.

23. R. Benedict, *Patterns of Culture* (Boston, New York, 1934), p. 151.

24. Ibid., pp. 148ff.

25. Helmuth, "Zum Verhalten des Menschen: Die Aggression," p. 269.

26. P. Weidkuhn, "Aggressivität und Normativität. Über die Vermittlerrolle der Religion zwischen Herrschaft und Freiheit. Ansätze zu einer kulturanthropologischen Theorie der sozialen Norm," *Anthropologie* 63/64 (1968/69): 361–94.

27. R.F. Fortune, "Arapesh Warfare," *American Anthropologist* 41 (1939): 22–41.

28. H. König, "Der Rechtsbruch und sein Ausgleich bei den Eskimos," *Anthropos* 20 (1925): 294.

29. R. Petersen, "Family Ownership and Right of Disposition in Sukkertoppen District, West Greenland," *Folk* 5 (1963): 270–81.

30. F. Boas, "The Eskimo of Baffin Land and Hudson Bay," *Bulletin of the American Museum of Natural History*, vol. 15 (1901–1907), p. 443.

31. K. Rasmussen, *Nye Mennesker* (Copenhagen, 1905), Introduction.

32. K. Rasmussen, *Report of the Fifth Thule Expedition*, vol. 9, *Intellectual Culture of the Copper Eskimos* (Copenhagen, 1932), p. 10. Quoted from Petersen, "Family Ownership and Right of Disposition in Sukkertoppen District, West Greenland."

33. Petersen, ibid., pp. 273ff. The references are to K. Lynge, *Kalâdlit oqalugtuait oqalualâvilo 1–3* (Nûk, 1938–39) and K. Rasmussen, *Myter og Saga fra Grønland* (Copenhagen, 1924).

34. Petersen, ibid., p. 278.

35. Ibid., pp. 280–81. Further information about the social structure and territoriality of the Central Eskimos can be found in D. Dumas, "Characteristics of Central Eskimo Band Structure," *Bulletin of the National Museum of Canada* no. 288 (1969), pp. 116–41.

36. H. W. Klutschak, *Als Eskimo unter Eskimos* (Vienna, 1881), p. 227.

37. Ibid., p. 150.

38. Ibid., p. 227.

39. G. Holm, "Ethnological Sketch of the Angmagssalik Eskimo," *Meddelelser om Grønland*, vol. 39 (1914); E. A. Hoebel, "Song Duels Among the Eskimo," in *Law and Warfare; Studies in the Anthropology of Conflict*, ed. P. Bohannan (New York, 1967), pp. 255–62.

40. K. Birket-Smith, *Die Eskimos* (Zurich, 1948); K. Rasmussen, *Myter og Saga fra Grønland*; idem, *Report of the Fifth Thule Expedition*, vol. 7, *Intellectual Culture of the Caribou Eskimos* (Copenhagen, 1930); idem, vol. 9, *Intellectual Culture of the Copper Eskimos*.

41. E. W. Nelson, *The Eskimo about Bering Strait, Eighteenth Annual Report of the Bureau of American Ethnology* (Washington, D.C., 1896/97), vol. 1.

42. Boas, "The Central Eskimo," pp. 424ff.

43. Service, *Primitive Social Organization*, p. 96.

44. C. M. Turnbull, *The Forest People* (London, New York, 1961); idem, *The Mbuti Pygmies: An Ethnographic Survey*, American Museum of Natural History Anthropological Papers no. 50 (1965).

45. C. M. Turnbull, "The Importance of Flux in Two Hunting Societies," in *Man the Hunter*, ed. R. B. Lee and I. DeVore (Chicago 1968), pp. 132–37.

46. P. Schebesta, *Die Bambuti-Pygmäen vom Ituri* (Brussels, 1941), 2:126.

47. Ibid., pp. 274ff.

48. M. G. Bicchieri, "The Differential Use of Identical Features of Physical Habitat in Connection with Exploitative Settlement and Community Patterns: The Bambuti," *Bulletin of the National Museum of Canada* no. 230 (1969) pp. 65–72. The reference is to idem, "A Study of the Ecology of Food-Gathering People: A Cross-Cultural Analysis of the Relationships of Environment, Technology, and Bio-Cultural Variability" (Doctoral dissertation, University of Minnesota, 1965).

49. Bicchieri, "The Differential Use of Identical Features of Physical Habitat in Connection with Exploitative Settlement and Community Patterns: The Bambuti," p. 69.

50. P. Schebesta, *Die Bambuti Pygmäen vom Ituri* (Brussels, 1948), 2:537.

51. P. Schumacher, *Kivu-Pymäen* (Brussels, 1950), vol. 2.

52. Woodburn, "Stability and Flexibility in Hadza Residential Groupings."

53. L. Kohl-Larsen, *Wildbeuter in Ostafrika: Die Tindiga, ein Jäger- und Sammlervolk* (Berlin, 1958), p. 101.

54. Ibid., p. 35. See also idem, *Auf den Spuren des Vormenschen* (Stuttgart, 1943).

55. I. Eibl-Eibesfeldt, *Die !Ko-Buschmanngesellschaft. Gruppenbindung und Aggressionskontrolle* (Munich, 1972).

56. Sahlins, "The Origin of Society."

57. Lee, "What Hunters Do for a Living," p. 31.

58. R. A. Rappaport, *Pigs for the Ancestors: Ritual in the Ecology of a New Guinea People* (New Haven, London, 1968); A. Ortiz, *The Tewa World: Space, Time, Being, and Becoming in a Pueblo Society* (Chicago, 1969); E. N. Wilmsen, "Interaction, Spacing Behavior, and the Organization of Hunting Bands," *Journal of Anthropological Research* 29 (1973): 1–31.

59. L. Marshall, "Sharing, Talking and Giving: Relief of Social Tensions among !Kung-Bushmen," *Africa* 31 (1961): 231–49; idem, "The !Kung-Bushmen of the Kalahari Desert," in *Peoples of Africa*, ed. J. L. Gibbs (New York, 1965), pp. 241–78.

60. W. Schmidbauer, "Zur Anthropologie der Aggression," *Dynamische Psychiatrie/Dynamic Psychiatry* 4 (1971): 36–50; idem, *Die sogenannte Aggression* (Hamburg, 1972); idem, "Territorialität und Aggression bei Jägern und Sammlern," *Anthropos* 68 (1973): 548–58.

61. R. B. Lee, "!Kung Spatial Organization: An Ecological and Historical Perspective," *Human Ecology* 1 (1972): 125–47.

62. S. Passarge, *Die Buschmänner der Kalahari* (Berlin, 1907), p. 31.

63. B. von Zastrow and H. Vedder, "Die Buschmänner," in *Das Eingeborenenrecht: Togo, Kamerun, Südwestafrica, die Südseekolonien*, ed. E. Schultz-Ewerth and L. Adam (Stuttgart, 1930), p. 425.

64. V. Lebzelter, *Eingeborenenkulturen von Süd- und Südwestafrika* (Leipzig, 1934), p. 21.

65. H. Vedder, "Die Buschmänner Südwestafrikas und ihre Weltanschauung," *South African Journal of Science* 24 (1937): 435.

66. H. Vedder, "Über die Vorgeschichte der Völkerschaften von Südwestafrika, pt. 1, Die Buschmänner," *Journal of the South-West Africa Scientific Society* 9 (1952/53): 49, 51.

67. [Oberleutnant] Trenk, "Die Buschleute der Namib, ihre Rechts- und Familienverhältnisse," *Mitteilungen der deutschen Schutzgebiete* (Berlin) 23 (1910): 168.

68. F. Brownlee, "The Social Organization of the !Kung (!Un) Bushmen of the North-Western Kalahari," *Africa* 14 (1943): 124–29;

J. H. Wilhelm, "Die !Kung-Buschleute," *Jahrbuch des Museums für Völkerkunde Leipzig* 12 (1953): 91–189. Wilhelm lived at Otjituo, South-West Africa, from 1914 to 1919. His very detailed notes were published in 1953, after his death.

69. Wilhelm, ibid., p. 156.

70. L. Marshall, "Marriage among the !Kung-Bushmen," *Africa* 29 (1959): 335–65.

71. L. Marshall, "The !Kung-Bushmen of the Kalahari Desert," in *Peoples of Africa*, ed. J. L. Gibbs (New York, 1965), p. 248.

72. P. von Tobias, "Bushman Hunter-Gatherers: A Study in Human Ecology," in *Ecological Studies in Southern Africa*, ed. D. H. S. Davis (The Hague, 1964). Reprinted (1968) in *Man in Adaptation: The Cultural Present*, ed. Y. A. Cohen (Chicago), pp. 196–208.

73. G. B. Silberbauer, "Socio-Ecology of the G/wi Bushmen" (Thesis, Department of Anthropology and Sociology, Monash University, 1973), p. 117. See also idem, "The G/wi Bushmen," in *Hunters and Gatherers Today: A Socioeconomic Study of Eleven Such Cultures in the Twentieth Century*, ed. M. G. Bicchieri (New York, 1972), pp. 271–326.

74. Silberbauer, "Socio-Ecology of the G/wi Bushmen," p. 212.

75. Lee, "!Kung Spatial Organization: An Ecological and Historical Perspective."

76. See L. Marshall, "!Kung-Bushmen Bands," *Africa* 30 (1960): 325–55.

77. H.-J. Heinz, "The Social Organization of the !Ko-Bushmen" (Master's thesis, University of South Africa, Johannesburg, 1966); idem, "Territoriality among the Bushmen in General and the !Ko in Particular," *Anthropos* 67 (1972): 407.

78. I. Eibl-Eibesfeldt, !Kung Buschleute (Kalahari)—Geschwisterrivalität, Mutter-Kind-Interaktionen," *Homo* 24 (1974): 252–60. For pictorial documentation, see idem, *Die !Ko-Buschmanngesellschaft*; and idem, *Der vorprogrammierte Mensch. Das Ererbte als bestimmender Faktor im menschlichen Verhalten* (Vienna, 1973).

79. M. J. Konner, "Aspects of the Developmental Ethology of a Foraging People," in *Ethological Studies of Child Behavior*, ed. N. Blurton-Jones (1972), pp. 285–304.

80. E. A. Tinbergen and N. Tinbergen, "Early Childhood Autism—An Ethological Approach," *Zeitschrift für Tierpsychologie*, Supplement 10 (1972).

81. See O. Koenig, *Kultur und Verhaltensforschung* (Munich, 1970).

82. H. Sbrzesny, "Die Spiele der !Ko-Buschleute unter besonderer Berücksichtigung ihrer socialisierenden und gruppenbindenden Funktion," *Monographien zur Humanethologie* 2 (Munich-Zurich, 1976).

83. R. B. Lee, "!Kung Bushmen Violence" (Paper read at a meeting of American Anthropological Association, November 1969); idem, "The !Kung Bushmen of Botswana," in *Hunters and Gatherers Today: A Socioeconomic Study of Eleven Such Cultures in the Twentieth Century*, ed. M. G. Bicchieri (New York, 1972), pp. 327–68.

84. H.-J. Heinz, "Conflicts, Tensions and Release of Tensions in a

Bushman Society," *Institute for the Study of Man in Africa Papers* no. 23 (1967), p. 6.

85. Eibl-Eibesfeldt, *Der vorprogrammierte Mensch.*

86. Documentary films HF 1 and 2; see bibliography, p. 285.

87. P. Germann, "Der Buschmannrevolver, ein Zaubergerät," *Jahrbuch des Museums für Völkerkunde Leipzig* 8 (1922): 51–56.

88. S. S. Dornan, *Pygmies and Bushmen of the Kalahari* (London, 1925).

89. Vedder, Über die Vorgeschichteder Völkerschaften von Südwestafrika. pt. 1, Die Buschmänner," *Journal of the South-West Africa Scientific Society* 9:45–56.

90. Wilhelm, "Die !Kung-Buschleute," p. 167.

91. H. Schindler, "Territorialität und Aggression: Eine Erwiderung." *Anthropos* 69 (1974): 275–78.

92. R. Gardner, in *Man the Hunter*, ed. R. B. Lee and I. DeVore (Chicago, 1968), p. 341.

93. M. Mead, *Growing up in New Guinea* (1930; reprinted in *From the South Seas* [New York, 1939]), p. 212.

94. M. Mead, *Sex and Temperament* (1935; reprinted in *From the South Seas*), p. 280.

95. Mead, *Growing up in New Guinea*, p. 239.

96. M. Mead, "The Role of the Individual in Samoan Culture," *Journal of the Royal Anthropological Institute* 58 (1928): 484.

97. A. Krämer, *Die Samoa Inseln. Entwurf einer Monographie unter besonderer Berücksichtigung Deutsch-Samoas*, vol. 1 (Stuttgart, 1902), p. 259.

98. M. Mead, *Coming of Age in Samoa* (1928; reprinted in *From the South Seas*), p. 105.

99. M. Harris, *The Rise of Anthropological Theory: A History of Theories of Culture* (New York, 1968).

100. D. Freeman, Letter to the Editor, *Current Anthropology*, 11 (1970): 66; idem, "The Evolutionary Theories of Charles Darwin and Herbert Spencer," *Current Anthropology* 15 (1974): 211–37.

101. E. Tobach, J. Gianutsos, H. R. Topoff, and C. G. Gross, *The Four Horsemen: Racism, Sexism, Militarism and Social Darwinism* (New York, 1974).

102. S. E. Luria, *Life: The Unfinished Experiment* (New York, 1973).

103. L. Berkowitz, *Aggression: A Social-Psychological Analysis* (New York, London, 1962), p. 4.

104. J. Rattner, *Aggression und menschliche Natur* (Olten, 1970), p. 34.

105. R. Denker, *Aufklärung über Aggression. Kant, Darwin, Freud, Lorenz* (Stuttgart, 1966), p. 95.

106. M. Lumsden, "The Instinct of Aggression: Science or Ideology?" *Futurum: Zeitschrift für Zukunftsforschung* 3 (1970): 408.

107. W. Lepenies and H. Nolte, *Kritik der Anthropologie* (Munich, 1971).

108. H. Selg, *Zur Aggression verdammt?* (Stuttgart, 1971); W. Hollitscher, ed., *Aggressionstrieb und Krieg* (Stuttgart, 1973); A. Schmidt-Mummendey and H. D. Schmidt, *Aggressives Verhalten* (Munich, 1971); A. Montagu, ed., *Man and Aggression* (New York, Toronto, 1968).

109. Schmidbauer, *Biologie und Ideologie: Kritik der Humanethologie* (Hamburg, 1973), p. 18.

110. B. F. Skinner, *Beyond Freedom and Dignity* (New York, 1971).

111. Eibl-Eibesfeldt, *Der vorprogrammierte Mensch.*

112. W. Michaelis, "Der Aggressions-'Trieb' im Streit der Zoologie und Psychologie," *Naturwissenschaftliche Rundschau* 27 (1974): 253.

113. Q. Wright, *A Study of War,* 2d ed. (Chicago, 1965), p. 22.

114. H. A. Bernatzik, *The Spirits of the Yellow Leaves* (London, 1958). Translation of (1941) *Die Geister der gelben Blätter.*

115. Wilhelm, "Die !Kung-Buschleute," pp. 159ff.

116. C. Strehlow, "Die Aranda- und Loritja-Stämme in Central-Australien," *Veröffentlichungen des Städtischen Völkermuseums Frankfurt/M.*, pt. 4, sec. 2, pp. 1–78.

117. Reported by F. E. Maning, *Old New Zealand: A Tale of the Good Old Times; and a History of the War in the North* (London, 1876), p. 147, quoted in A. P. Vayda, "Maoris and Muskets in New Zealand: Disruption of a War System," *Political Science Quarterly* 85 (1970): 560–84.

118. C. Lumholtz, *Among Cannibals. An Account of Four Years' Travels in Australia and of Camp Life with the Aborigines of Queensland* (London, 1889), pp. 123–27.

119. G. Landtman, *The Kiwai Papuans of British New Guinea* (London, 1927).

120. Conditions among the Duum Dani have been described by P. Matthiessen, *Under the Mountain Wall. A Chronicle of Two Seasons in the Stone Age* (New York, 1962); and R. Gardner and K. G. Heider, *Gardens of War: Life and Death in the New Guinea Stone Age* (New York, 1968.)

121. Gardner and Heider, p. 138.

122. Ibid., p. 139.

123. Ibid., p. 142.

124. J. Layard, *Stone Men of Malekula* (London, 1942). Other examples of warlike encounters between primitives are to be found in L. Frobenius, *Weltgeschichte des Krieges* (Jena, 1903); W. E. Mühlmann, *Krieg und Frieden. Ein Leitfaden der politischen Ethnologie* (Heidelberg, 1940); H. H. Turney-High, *Primitive War: Its Practice and Concepts* (Columbia, S.C., 1949); P. Bohannan, ed., *Law and Warfare: Studies in the Anthropology of Conflict* (Garden City, N.Y., 1967); M. Fried, M. Harris, and R. Murphy, *War: The Anthropology of Armed Conflict and Aggression* (Garden City, N.Y., 1968). I shall deal in a separate chapter with the ways in which they prevent or end warfare or conduct it without bloodshed.

125. K. F. Koch, "Cannibalistic Revenge in Jalé Society," *Natural History* 79, 2 (1970): 40–51.

126. E. Volhard, *Kannibalismus* (Stuttgart, 1939; reprint ed., New York, 1968).

127. M. D. Dornstreich and G. E. B. Morren, "Does New Guinea Cannibalism Have Nutritional Value?" *Human Ecology* 2 (1974): 1–12.

128. M. N. Walsh, "Psychic Factors in the Causation of Recurrent Mass Homicide," in *War and the Human Race*, ed. idem, (New York, 1971).

129. E. Fromm, *The Anatomy of Human Destructiveness* (New York, London, 1973).

130. C. R. Hallpike, "Functionalist Interpretations of Primitive Warfare," *Man* 8 (1973): 451–70.

131. G. Kramer, "Macht die Natur Konstruktionsfehler?" *Wilhelmshavener Vorträge, Schriftenreihe der Nordwestdeutschen Universitätsgesellschaft* 1 (1949): 1–19.

132. Fromm, *The Anatomy of Human Destructiveness*, p. 19.

133. R. S. Bigelow, *The Dawn Warriors: Man's Evolution Toward Peace* (London, 1970); idem, "Relevance of Ethology to Human Aggressiveness," *International Social Science Journal* 23 (1971): 19–26.

134. Since the first publication of this book, an interesting article dealing with the selective advantage of warfare has appeared: W. H. Durham, "Resource Competition and Human Aggression, pt. 1, A Review of Primitive War," *Quarterly Review of Biology* 51 (1976): 385–415.

135. N. A. Chagnon, "Yanomamö Social Organization and Warfare," in *War: The Anthropology of Armed Conflict and Aggression*, ed. M. Fried, M. Harris, and R. Murphy (Garden City, N.Y., 1968), pp. 109–59; idem, *Yanomamö: The Fierce People* (New York, 1968).

136. Wright, *A Study of War*, p. 76.

137. Rappaport, *Pigs for the Ancestors*.

138. Wright, *A Study of War*; A. P. Vayda, "Maori Warfare," in *Polynesian Society Maori Monographs* no. 2 (Wellington, 1960); idem, "Expansion and Warfare Among Swidden Agriculturalists," *American Anthropologist* 63 (1961): 346–58; idem, "Research on the Functions of Primitive War," in *Peace Research Society International Papers*, vol. 7 (1967); idem, "Hypotheses About Functions of War," in *War: The Anthropology of Armed Conflict and Aggression*, ed. M. Fried, M. Harris, and R. Murphy, pp. 85–91, with further literature in the last.

139. R. V. Morey, Jr., and J. P. Marwitt, "Ecology, Economy and Warfare in Lowland South America" (Paper read at 9th International Congress of Anthropological and Ethnological Sciences, August–September 1973, Chicago).

140. E. F. Castetter and W. H. Bell, *Yuman Indian Agriculture* (Albuquerque, 1951); E. E. Graham, "Yuman Warfare: An Analysis of Ecological Factors from Ethnohistorical Sources" (Paper read at 9th International Congress of Anthropological and Ethnological Sciences, August–September 1973, Chicago).

141. Vayda, "Maoris and Muskets in New Zealand: Disruption of a War System."

142. Layard, *Stone Men of Malekula*, p. 603.

CHAPTER 7 ON THE WAY TO PEACE

1. H. Portisch, *Friede durch Angst: Augenzeuge in den Arsenalen des Atomkrieges* (Vienna, 1970), p. 145.

2. W. L. Shirer, *The Rise and Fall of the Third Reich* (New York, 1960), pp. 598–99.

3. K. Oberhuber, *Die Kultur des alten Orients. Handbuch der Kulturgeschichte* (Frankfurt, 1972), pp. 282–87.

4. L. P. Edwards, *The Natural History of Revolution* (Chicago, 1927; reprint ed. 1970), pp. 178ff.

5. Reported in E. Biocca, *Yanoáma: The Narrative of a White Girl Kidnapped by Amazonian Indians* (New York, 1970).

6. T. Hahn, "Die Buschmänner, pt. 3," *Globus: Zeitschrift für Länder- und Völkerkunde* 18 (1870): 102–105.

7. W. Wickler, *Die Biologie der Zehn Gebote* (Munich, 1971).

8. M. J. Leahy and M. Crain, *The Land that Time Forgot* (London, 1937).

9. See I. Eibl-Eibesfeldt, *Der vorprogrammierte Mensch. Das Ererbte als bestimmender Faktor in menschlichen Verhalten* (Vienna, 1973); T. K. Pitcairn and M. Schleidt, "Dance and Decision: An Analysis of a Courtship Dance of the Medlpa, New Guinea," *Behaviour* 58 (1976): 298–316.

10. See Eibl-Eibesfeldt, *Der vorprogrammierte Mensch.*

11. W. L. Warner, "Murngin Warfare," *Oceania* 1 (1930): 457–94.

12. R. A. Rappaport, *Pigs for the Ancestors: Ritual in the Ecology of a New Guinea People* (New Haven, London, 1968).

13. G. F. Vicedom, "Ein neuentdecktes Volk in Neuguinea," *Archiv für Anthropologie*, n.s. 24 (1937): 11–44.

14. A. Strathern, *The Rope of Moka: Big-Men and Ceremonial Exchange in Mount Hagen, New Guinea* (London, New York, 1971), p. 77.

15. G. F. Vicedom and H. Tischner, *Die Kultur der Hagenbergstämme*, vol. 1, *Die Mbowamb* (Hamburg, 1943/48).

16. Strathern, *The Rope of Moka*, p. 54.

17. K. F. Koch, *War and Peace in Jalémó: The Management of Conflict in Highland New Guinea* (Cambridge, Mass., 1974), p. 83.

18. G. Landtman, *The Kiwai Papuans of British New Guinea* (London, 1927).

19. Vicedom and Tischner, *Die Kultur der Hagenbergstämme*, vol. 1.

20. I. Eibl-Eibesfeldt, "Phylogenetic and Cultural Adaptation in Human Behavior," in *Animal Models in Human Psychobiology*, ed. G. Serban and A. Kling (New York, 1976), pp. 77–98.

21. Biocca, *Yanoáma.*

22. A. R. Radcliffe-Brown, *The Andaman Islanders: A Study in Social Anthropology*, 2d ed. (Cambridge, New York, 1933).

23. See also H. Codere, *Fighting with Property*. American Ethnological Society Monograph no. 18; M. D. Sahlins, *Stone Age Economics* (Chicago, 1972); M. Mauss, *The Gift: Forms and Functions of Exchange in Archaic Societies* (Glencoe, Ill., London, 1954).

24. Eibl-Eibesfeldt, *Der vorprogrammierte Mensch*.

25. L. R. Hiatt, *Kinship and Conflict: A Study of an Aboriginal Community in Northern Arnhem Land* (Canberra, 1965).

26. Mauss, *The Gift*; Sahlins, *Stone Age Economics*.

27. W. Wickler, *Die Biologie der Zehn Gebote* (Munich, 1971); H. Kummer, W. Götz, and W. Angst, "Triadic Differentiation: An Inhibitory Process Protecting Pair Bonds in Baboons," *Behaviour* 49 (1974): 62–87.

28. P. Bamm, *Eines Menschen Zeit* (Munich, 1972), pp. 41–43.

29. H. U. Wintsch, "Erziehung zur Friedfertigkeit," in *Der Mythos vom Aggressionstrieb*, ed. A. Plack (Munich, 1973), p. 288.

30. A. Mitscherlich, *Die Idee des Friedens und die menschliche Aggressivität* (Frankfurt, 1969), p. 108.

31. M. Mead, "Alternatives to War," in *War: The Anthropology of Armed Conflict and Aggresssion*, ed. M. Fried, M. Harris, and R. Murphy (Garden City, N.Y., 1968), pp. 222–23.

32. A. Gehlen, *Moral und Hypermoral* (Frankfurt, 1969), p. 121.

33. K. Lorenz, *Behind the Mirror: A Search for a Natural History of Knowledge* (New York, 1977). Translation of (1973) *Die Rückseite des Spiegels. Versuch einer Naturgeschichte menschlichen Erkennens*.

34. T. Dobzhansky, "Evolution and Man's Self-Image," in *The Quest for Man*, ed. V. Goodall (London, New York, 1975).

35. I. Fetscher, *Modelle der Friedenssicherung* (Munich, 1972); see additional literature there.

36. R. G. Sipes, "War, Sports and Aggression: An Empirical Test of Two Rival Theories," *American Anthropologist* 75 (1973): 64–86.

37. H. Marcuse, *Triebstruktur und Gesellschaft* (Frankfurt, 1969).

38. C. Meves, *Manipulierte Masslosigkeit* (Freiburg, 1971).

39. B. Hassenstein, "Wesensverschiedene Formen menschlicher Aggressivität in der Sicht der Verhaltensforschung," *Universitas* 28 (1973): 287–95.

40. Wintsch, "Erziehung zur Friedfertigkeit."

41. M. Mead, "How Do Children Learn to Govern Their Own Violent Impulses?" *American Journal of Orthopsychiatry* 39 (1969): 229.

42. R. K. Dentan, *The Semai: A Nonviolent People of Malaya* (New York, 1968), p. 59.

Bibliography

ALBRECHT, H., and DUNNETT, S. C. 1971. *Chimpanzees in Western Africa.* Munich.

ARDREY, R. 1966. *The Territorial Imperative.* New York.

ARGYLE, M., and COOK, M. 1976. *Gaze and Mutual Gaze.* Cambridge.

BAGSHAWE, F. J. 1924/25. "The Peoples of the Happy Valley (East Africa)." *Journal of African Society* 24: 25–33, 117–30, 219–27, 328–47.

BAKER, J. W., and SCHAIE, K. W. 1969. "Effects of Aggression 'Alone' or 'with Another' on Physiological and Psychological Arousal." *Journal of Personality and Social Psychology* 12: 80–96.

BALL, W., and TRONICK, F. 1966. "Infant Responses to Impending Collision: Optical and Real." *Science* 151: 818–820.

BAMM, P. 1972. *Eines Menschen Zeit.* Munich.

BANDURA, A. 1973. *Aggression: A Social Learning Analysis.* New York.

———, and WALTERS, R. H. 1963. *Social Learning and Personality Development.* New York.

BARON, R. A. 1974. "Sexual Arousal and Physical Aggression: The Inhibiting Influence of 'Cheesecake' and Nudes." *Bulletin of the Psychonomic Society* 3: 337–39.

———. 1974. "The Aggression-Inhibiting Influence of Heightened Sexual Arousal." *Journal of Personality and Social Psychology* 30: 318–22.

———, and BALL, R. L. 1974. "The Aggression-Inhibiting Influence of Non-Hostile Humor." *Journal of Experimental Social Psychology* 10: 23–33.

BASEDOW, H. 1906. "Anthropological Notes on the Western Coastal Tribes of the Northern Territory of South Australia." *Transactions of the Royal Society of South Australia* 31: 1–62.

BEHRENS, H. 1974. "Historische Bewegkräfte im Neolithikum Mitteleuropas." *Archaeologia Austriaca* 55: 91–94.

———. 1975. "Der Kampf als historische Bewegkraft in der Steinzeit." *Jahresschrift für mitteldeutsche Vorgeschichte* 59.

BENEDICT, R. 1934. *Patterns of Culture.* Boston, New York.

BERKOWITZ, L. 1962. *Aggression: A Social-Psychological Analysis.* New York, London.

———. 1970. "Aggressive Humor as a Stimulus to Aggressive Responses." *Journal of Personality and Social Psychology* 16: 710–17.

———, CORWIN, R., and HEIRONIMUS, M. 1963. "Film Violence and Subsequent Aggressive Tendencies." *Public Opinion Quarterly* 27: 217–29.

BERNATZIK, H. A. 1958. *The Spirits of the Yellow Leaves*. London. Translation of (1941) *Die Geister der gelben Blätter*.

BICCHIERI, M. G. 1965. "A Study of the Ecology of Food-Gathering People: A Cross-Cultural Analysis of the Relationships of Environment, Technology, and Bio-Cultural Variability." Doctoral dissertation, University of Minnesota.

———. 1969. "The Differential Use of Identical Features of Physical Habitat in Connection with Exploitative Settlement and Community Patterns: The Bāmbuti." *Bulletin of the National Museum of Canada* no. 230: 65–72.

———. 1972. *Hunters and Gatherers Today: A Socioeconomic Study of Eleven Such Cultures in the Twentieth Century*. New York.

BIGELOW, R. S. 1970. *The Dawn Warriors: Man's Evolution Toward Peace*. London.

———. 1971. "Relevance of Ethology to Human Aggressiveness." *International Social Science Journal* 23: 19–26.

———. 1972. "The Evolution of Cooperation, Aggression and Self-Control." In *Nebraska Symposium on Motivation*, ed. J. K. Cole and D. D. Jensen, pp. 1–57. Lincoln.

BIOCCA, E. 1970. *Yanoáma: The Narrative of a White Girl Kidnapped by Amazonian Indians*. New York.

BIRKET-SMITH, K. 1948. *Die Eskimos*. Zurich.

BLEEK, D. F. 1930. *Rock-Paintings in South Africa*. London.

BOAS, F. 1888. "The Central Eskimo." *Sixth Annual Report of the Bureau of American Ethnology, 1884–85:* 399–669. Washington, D.C.

———. 1901–1907. "The Eskimo of Baffin Land and Hudson Bay." *Bulletin of the American Museum of Natural History*, vol. 15, pts. 1 and 2.

BOHANNAN, P. ed. 1967. *Law and Warfare: Studies in the Anthropology of Conflict*. Garden City, N.Y.

BOWER, T. G. 1966. "Slant Perception and Shape Constance in Infants," *Science* 151: 832–34.

———. 1971. "The Object in the World of the Infant." *Scientific American* 255: 30–38.

BRAMEL, D., TAUB, B., and BLUM, B. 1968. "An Observer's Reaction to the Suffering of His Enemy." *Journal of Personality and Social Psychology* 8: 384–92.

BROWNLEE, F. 1943. "The Social Organization of the !Kung (!Un) Bushmen of the North-Western Kalahari." *Africa* 14: 124–29.

BULLOCK, T. H., and HORRIDGE, G. A. 1965. *Structure and Function in the Nervous System of Invertebrates*, vols. 1 and 2. San Francisco.

BUSS, A. H. 1961. *The Psychology of Aggression*. New York.

BYGOTT, J. D. 1972. "Cannibalism among Wild Chimpanzees." *Nature* 238: 410–11.

CAIRNS, R. B. 1972. "Fighting and Punishment from a Developmental Perspective." In *Nebraska Symposium on Motivation*, ed. J. K. Cole and D. D. Jensen, pp. 59–124. Lincoln.

CARPENTER, C. R. 1940. *A Field Study in Siam of the Behavior and Social Relations of the Gibbon*. Comparative Psychology Monographs no. 16. Reprinted (1964) in *Naturalistic Behavior of Nonhuman Primates*, pp. 145–271. University Park, Pa.

CASTETTER, E. F., and BELL, W. H. 1951. *Yuman Indian Agriculture*. Albuquerque.

CHAGNON, N. A. 1968. *Yanomamö: The Fierce People*. New York.

———. 1968. "Yanomamö Social Organization and Warfare." In *War: The Anthropology of Armed Conflict and Aggression*, ed. M. Fried, M. Harris, and R. Murphy, pp. 109–59. Garden City, N.Y.

CHANCE, M. R. A. 1962. "An Interpretation of Some Agonistic Postures: The Role of 'Cut-off' Acts and Postures." *Symposia of the Zoological Society of London* 8: 71–89.

———. 1967. "Attention Structure as the Basis of Primate Rank Orders." *Man* 2: 503–18.

CLAUSEWITZ, K. VON. 1937. *Vom Kriege*. Ed. K. Linnebach. Berlin.

CODERE, H. 1950. *Fighting with Property: A Study of Kwakiutl Potlatching and Warfare*. American Ethnological Society Monograph no. 18.

CONRADT, R. 1973. "Intergruppenaggression–ein artspezifisches Merkmal des Menschen." *Universitas* 28: 1013–18.

CORNING, P. A. 1973. "Comparative Survival Strategies: An Approach to Social and Political Analysis." Paper read at annual meeting of American Political Science Association, September 1973, New Orleans.

COSS, R. G. 1972. "Eye-Like Schemata: Their Effect on Behavior." Dissertation, University of Reading.

COURCHESNE, E., and BARLOW, G. W. 1971. "Effect of Isolation on Components of Aggressive and Other Behavior in the Hermit Crab *Pagurus samuelis*." *Zeitschrift für vergleichende Physiologie* 75: 32–48.

DAMAS, D. 1969. "Characteristics of Central Eskimo Band Structure." *Bulletin of the National Museum of Canada* no. 228: 116–41.

DANN, H. D. 1972. *Aggression und Leistung*. Stuttgart.

DART, R. A. 1949. "The Bone-bludgeon Hunting Technique of Australopithecus." *South African Science* 2: 150–52.

———. 1953. "The Predatory Transition from Ape to Man." *International Anthropological and Linguistic Review* 1: 201–18.

DARWIN, C. 1859. *The Origin of Species*. London.

———. 1871. *The Descent of Man*. London.

———. 1872. *The Expression of Emotion in Man and Animals*. London.

DAWKINS, R. 1976. *The Selfish Gene*. Oxford.

DEAG, J. M., and CROOK, J. H. 1971. "Social Behaviour and 'Agonistic Buffering' in the Wild Barbary Macaque *Macaca sylvana*." *Folia Primatologica* 15: 183–200.

DELGADO, J. M. R. 1966. "Evoking and Inhibiting Aggressive Behavior by Radiostimulation in Monkey Colonies." *American Zoologist* 6: 669–81.

———. 1969. "Radiostimulation of the Brain in Primates and in Man." *Anesthesia and Analgesia Current Researches* 48: 529–43.

DENKER, R. 1966. *Aufklärung über Aggression. Kant, Darwin, Freud, Lorenz.* Stuttgart.

DENTAN, R. K. 1968. *The Semai: A Nonviolent People of Malaya.* New York.

DETHIER, V. G. 1957. "Communication by Insects: Physiology of Dancing." *Science* 125: 331–36.

———, and BODENSTEIN, D. 1958. "Hunger in the Butterfly." *Zeitschrift für Tierpsychologie* 15: 129–40.

DE VORE, I. 1971. "The Evolution of Human Society." In *Man and Beast: Comparative Social Behavior*, Smithsonian Annual 3, ed. J. F. Eisenberg and W. S. Dillon, pp. 297–311. Washington, D.C.

DIVALE, W. T. 1972. "System Population Control in the Middle and Upper Palaeolithic: Inferences Based on Contemporary Hunter-Gatherers." *World Archaeology* 4: 222–43.

DOBZHANSKY, T. 1975. "Evolution and Man's Self-Image." In *The Quest for Man*, ed. V. Goodall, pp. 189–220. London.

DOLLARD, J., DOOB, L. W., MILLER, N. E., MOWRER, O. H., and SEARS, R. R. 1939. *Frustration and Aggression.* New Haven.

DOOB, A. N., and WOOD, L. 1972. "Catharsis and Aggression: The Effects of Annoyance and Retaliation on Aggressive Behavior." *Journal of Personality and Social Psychology* 28: 156–62.

DORNAN, S. S. 1925. *Pygmies and Bushmen of the Kalahari.* London.

DORNSTREICH, M. D., and MORREN, G. E. B. 1974. "Does New Guinea Cannibalism Have Nutritional Value?" *Human Ecology* 2: 1–12.

DUMAS, D. 1969. "Characteristics of Central Eskimo Band Structure." *Bulletin of the National Museum of Canada* 288: 116–41.

DURHAM, W. H. 1976. "Resource Competition and Human Aggression, pt. 1, A Review of Primitive War." *Quarterly Review of Biology* 51: 385–415.

DWORKIN, E. S., and EFRAN, J. S. 1967. "The Angered: Their Susceptibility to Varieties of Humor." *Journal of Personality and Social Psychology* 6: 233–36.

EDNEY, J. J. 1972. "Property, Possession and Permanence: A Field Study in Human Territoriality." *Journal of Applied Social Psychology* 2, 275–282.

———, and JORDAN-EDNEY, N. L. 1974. "Territorial Spacing on a Beach." *Sociometry* 37: 92–104.

EDWARDS, L. P. 1927. *The Natural History of Revolution.* Chicago. Reprinted (1970).

EIBL-EIBESFELDT, I. 1953. "Zur Ethologie des Hamsters (*Cricetus cricetus* L.)." *Zeitschrift für Tierpsychologie* 10: 204–54.

———. 1963. "Angeborenes und Erworbenes im Verhalten einiger Säuger." *Zeitschrift für Tierpsychologie* 20: 705–54.

———. 1967. *Land of a Thousand Atolls.* London, New York. Translation of (1964) *Im Reich der tausend Atolle.*

———. 1970. "Männliche und weibliche Schutzamulette im modernen Japan." *Homo* 21: 175–88.

———. 1972. *Die !Ko-Buschmanngesellschaft. Gruppenbindung und Aggressionskontrolle*. Munich.

———. 1973. *Der vorprogrammierte Mensch. Das Ererbte als bestimmender Faktor im menschlichen Verhalten*. Vienna.

———. 1974. *Love and Hate: The Natural History of Behavior Patterns*. New York, London. Translation of (1970) *Liebe und Hass. Zur Naturgeschichte elementarer Verhaltensweisen*.

———. 1974. !Kung-Buschleute (Kalahari)–Geschwisterrivalität, Mutter-Kind-Interaktionen. *Homo* 24: 252–60.

———. 1974. "Medlpa (Mbowamb)–Neuguinea–Werberitual (Amb Kanànt)." *Homo* 25: 274–84.

———. 1975. *Ethology: The Biology of Behavior*. New York, London. Translation of (1974) *Grundriss der vergleichenden Verhaltensforschung*, 4th ed.

———. 1976. "Phylogenetic and Cultural Adaptation in Human Behavior." In *Animal Models in Human Psychobiology*, ed. G. Serban and A. Kling, pp. 77–98. New York.

———, and HASS, H. 1966. "Zum Projekt einer ethologisch orientierten Untersuchung menschlichen Verhaltens." *Mitteilungen der Max-Planck-Gesellschaft* 6: 383–96.

———. In press. "Rituals and Ritualisation from a Biological Perspective." In *Human Ethology, Claims and Limits of a New Discipline*, ed. M. von Cranach, K. Foppa, W. Lepenies, and D. Ploog. Cambridge.

———. In press. "Strategies of Social Interaction." In *Applications and Cross-Cultural Implications*, ed. R. Plutchik and H. Kellerman. New York.

———. In press. "Human Ethology: Concepts and Implications for the Sciences of Man." *Behavior and Brain Sciences*.

———, and WICKLER, W. 1968. "Die ethologische Deutung einiger Wächterfiguren auf Bali." *Zeitschrift für Tierpsychologie* 25: 719–26.

ERIKSON, E. H. 1966. "Ontogeny of Ritualization in Man." *Philosophical Transactions of the Royal Society of London*, ser. B, 251: 337–49.

ESSER, A. H. 1970. "Interactional Hierarchy and Power Structure on a Psychiatric Ward." In *Behaviour Studies in Psychiatry*, ed. S. J. Hutt and C. Hutt, pp. 25–59. Oxford, New York.

EULER, H. A. 1972. "Der Effekt von aggressionsabhängiger Strafreizung (Elektroschock) auf das Kampfverhalten von Leghorn Hähnen." Paper read at 28. Kongress der Deutschen Gesellschaft für Psychologie, October 1972. *Gruppendynamik* 3: 311–18.

FELIPE, N. J., and SOMMER, R. 1966. "Invasions of Personal Space." *Social Problems* 14: 206–14.

FESHBACH, S. 1961 "The Stimulating versus Cathartic Effects of a Vicarious Aggressive Activity." *Journal of Abnormal and Social Psychology* 63: 381–85.

———. 1964. "The Function of Aggression and the Regulation of Aggressive Drive." *Psychological Review* 71: 257–72.

———, and SINGER, R. 1971. *Television and Aggression*. San Francisco.

FETSCHER, I. 1972. *Modelle der Friedenssicherung*. Munich.

FORTUNE, R. F. 1939. "Arapesh Warfare." *American Anthropologist* 41: 22–41.

FOSSEY, D. 1971. "More Years with Mountain Gorillas." *National Geographic Magazine* 140: 574–85.

FRANCK, D., and WILHELMI, U. 1973. "Veränderungen der aggressiven Handlungsbereitschaft männlicher Schwertträger, *Xiphophorus helleri*, nach sozialer Isolation." *Experientia* 29: 896–97.

FREEMAN, D. 1970. Letter to the Editor. *Current Anthropology* 11: 66.

———. 1971. "Aggression: Instinct or Symptom?" *Australian & New Zealand Journal of Psychiatry*, 5: 66–73.

———. 1974. "The Evolutionary Theories of Charles Darwin and Herbert Spencer." *Current Anthropology* 15: 211–37.

FREUD, S. 1913. "Totem and Tabu II: Das Tabu und die Ambivalenz der Gefühlsregnungen." *Imago* 1: 213–27 and 301–33. Reprinted in *Condition Humana* (Frankfurt, 1974), S. Freud Studien Ausgabe, Vol. 9.

FRIED, M., HARRIS, M., and MURPHY, R., eds. 1968. *War: The Anthropology of Armed Conflict and Aggression*. Garden City, N.Y.

FROBENIUS, L. 1903. *Weltgeschichte des Krieges*. Jena.

FROMM, E. 1973. *The Anatomy of Human Destructiveness*. New York, London.

———. 1974. "Lieber fliehen als kämpfen." *Bild der Wissenschaft* 10: 52–58.

GARDNER, R., and HEIDER, K. G. 1968. *Gardens of War: Life and Death in the New Guinea Stone Age*. New York.

GEEN, R. G., STONNER, D., and SHOPE, G. L. 1975. "The Facilitation of Aggression by Aggression: Evidence against the Catharsis Hypothesis." *Journal of Personality and Social Psychology* 31: 721–26.

GEHLEN, A. 1969. *Moral und Hypermoral*. Frankfurt.

GERMANN, P. 1922. "Der Buschmannrevolver, ein Zaubergerät." *Jahrbuch des Museums für Völkerkunde Leipzig* 8: 51–56.

GIBBS, F. A. 1951. "Ictal and Non-Ictal Psychiatric Disorders in Temporal Lobe Epilepsy." *Journal of Nervous and Mental Disease* 113: 522–28.

GLEICHEN-RUSSWURM, A. von 1930. *Kultur- und Sittengeschichte aller Zeiten und Völker*. Vols. 9/10, *Kultur und Geist der Renaissance*, vol. 9, *Das Jahrhundert des Europäischen Humanismus*. Hamburg.

GODELIER, M. 1978. "Territory and Property in Primitive Society." *Social Science Information* 17: 399–426.

GOLDENBOGEN, I. 1977. "Über den Einfluss sozialer Isolation auf die aggressive Handlungsbereitschaft von Xiphophorus helleri (Poeciliidae) und *Haplochromis burtoni* (Cichlidae)." *Zeitschrift für Tierpsychologie* 44: 25–44.

GOODALL, J. 1965. "Chimpanzees of the Gombe Stream Reserve." In *Pri-*

mate Behavior: Field Studies of Monkeys and Apes, ed. I. DeVore, pp. 425–73. New York.

See also Lawick-Goodall, J. van.

GRAHAM, E. E. 1973. "Yuman Warfare: An Analysis of Ecological Factors from Ethnohistorical Sources." Paper read at 9th International Congress of Anthropological and Ethnological Sciences, August-September 1973, Chicago.

GRASTYAN, E. 1974. "Emotion." In *Encyclopaedia Britannica*. 15th ed. *Macropaedia*, 6: 757–66.

GRAUMANN, C. F. 1969. *Einführung in die Psychologie*. Frankfurt.

HACKER, F. 1971. *Aggression*. Vienna.

HAHN, T. 1870. "Die Buschmänner, pt. 3," *Globus: Zeitschrift für Länder- und Völkerkunde* 18: 102–105.

HALLPIKE, C. R. 1973. "Functionalist Interpretations of Primitive Warfare." *Man* 8: 451–70.

HAMILTON, W. D. 1964. "The Genetical Evolution of Social Behavior." *Journal of Theoretical Biology* 7: 1–52.

HARRIS, M. 1968. *The Rise of Anthropological Theory: A History of Theories of Culture*. New York.

———. 1977. *Cannibals and Kings: The Origins of Cultures*. New York.

HASS, H. 1970. *Das Energon*. Vienna.

HASSENSTEIN, B. 1973. "Wesensverschiedene Formen menschlicher Aggressivität in der Sicht der Verhaltensforschung." *Universitas* 28: 287–95.

———. 1973. *Verhaltensbiologie des Kindes*. Munich.

HEBB, D. O. 1946. "On the Nature of Fear." *Psychological Review* 53: 259–76.

HEESCHEN, V., SCHIEFENHOVEL, W., and EIBL-EIBESFELDT, I. In press. "Requesting, Giving and Taking: The Relationship Between Verbal and Nonverbal Behavior in the Speech Community of the Eipo, Irian Jaya (West New Guinea)." In *Verbal and Nonverbal Communication*, ed. R. Key. The Hague.

HEILIGENBERG, W. 1964. "Ein Versuch zur ganzheitsbezogenen Analyse des Instinktverhaltens eines Fisches (*Pelmatochromis subocellatus kribensis, Boul., Cichlidae*)." *Zeitschrift für Tierpsychologie* 21: 1–52.

———, *and* KRAMER, U. 1972. "Aggressiveness as a Function of External Stimulation." *Journal of Comparative Physiology* 77: 332–40.

HEINZ, H.-J. 1966. "The Social Organization of the !Ko Bushmen." Master's thesis, University of South Africa, Johannesburg.

———. 1967. "Conflicts, Tensions and Release of Tensions in a Bushman Society." *Institute for the Study of Man in Africa Papers* no. 23, pp. 2–21.

———. 1972. "Territoriality Among the Bushmen in General and the !Kc in Particular." *Anthropos* 67: 405–16.

———. In press. "The Nexus System among the !Ko." *Anthropos*.

———, and MAGUIRE, B. 1974. "The Ethno-Biology of the !Ko-Bushmen." Botswana Society Occasional Paper no. 1. Gaborone.

HELMUTH, H. 1967. "Zum Verhalten des Menschen: Die Aggression." *Zeitschrift für Ethnologie* 92: 265–73.

HESS, E. H. 1973. *Imprinting: Early Experience and the Developmental Psychobiology of Attachment.* New York, London.

HIATT, L. R. 1965. *Kinship and Conflict: A Study of an Aboriginal Community in Northern Arnhem Land.* Canberra.

HICKS, D. J. 1965. "Imitation and Retention of Film-Mediated Aggressive Peer and Adult Models." *Journal of Personality and Social Psychology* 2: 97–100.

HINDE, R. A. 1966. *Animal Behavior: A Synthesis of Ethology and Comparative Psychology.* New York, London.

HOBHOUSE, L. T. 1956. "The Simplest Peoples, pt. 2, Peace and Order among the Simplest Peoples," *British Journal of Sociology* 7: 96–119.

HOEBEL, E. A. 1967. "Song Duels Among the Eskimo." In *Law and Warfare: Studies in the Anthropology of Conflict*, ed. P. Bohannan, pp. 255–62. Garden City, N.Y.

HÖRMANN, L. VON. 1877. *Tiroler Volkstypen.* Vienna.

———. 1909. *Tiroler Volksleben.* Stuttgart.

HOKANSON, J. E., and SHETLER, S. 1961. "The Effect of Overt Aggression on Physiological Tension Level." *Journal of Abnormal and Social Psychology* 63: 446–48.

———. 1970. "Psychophysiological Evaluation of the Catharsis Hypothesis." In *The Dynamic of Aggression*, ed. E. I. Megargee and J. E. Hokanson. New York.

HOLD, B. 1974. "Rangordnungsverhalten bei Vorschulkindern. Eine vorläufige Mitteilung." *Homo* 25: 252–67.

———. 1974. "Rank and Behaviour: An Ethological Study of Preschool Children." *Homo* 28: 154–88.

HOLLITSCHER, W. ed. 1973. *Aggressionstrieb und Krieg.* Stuttgart.

HOLM, G. 1914. "Ethnological Sketch of the Angmagssalik Eskimo." *Meddelelser om Grønland,* vol. 39.

HOLST, E. von. 1939. "Die relative Koordination als Phänomen und als Methode zentralnervöser Funktionsanalyse." *Ergebnisse der Physiologie* 42: 228–306. Reprinted (1969) in idem, *Zur Verhaltensphysiologie bei Tieren und Menschen. Gesammelte Abhandlungen,* vol. 1. Munich.

———. 1961. "Probleme der modernen Instinktforschung," reprinted (1969) in idem, *Zur Verhaltensphysiologie bei Tieren und Menschen. Gesammelte Abhandlungen,* vol. 1. Munich.

———, and SAINT-PAUL, U. von. 1960. "Vom Wirkungsgefüge der Triebe." *Die Naturwissenschaften* 18: 409–22.

HOOFF, J. A. R. A. M. van. 1971. *Aspecten van het Sociale Gedrag en de Communicatie Bij Humane en Hogere Niet-Humane Primaten (Aspects of Social Behavior and Communication in Human and Higher Non-Human Primates).* Rotterdam.

HORN, K. 1974. "Die humanwissenschaftliche Relevanz der Ethologie." In *Kritik der Verhaltensforschung*, ed. G. Roth, pp. 190–221. Munich.

HUMM, R. J. 1958. "Der Mann, der die Tiersprache versteht." *Die Weltwoche* (Zurich), February 3.
HUNSPERGER, R. W. 1954. "Reizversuche in periventrikulären Grau des Mittel- und Zwischenhirns" (Film). *Helvetica Physiologica Acta* 12: C4–C6.
IMMELMANN, K. 1970. "Zur ökologischen Bedeutung prägungsbedingter Isolationsmechanismen." In *Verhandlungen der Deutschen Zoologischen Gesellschaft, 64. Jahresversammlung 1970*, ed. W. Rathmayer, pp. 304–14. Stuttgart.
ITANI, J., *and* SUZUKI, A. 1967. "The Social Unit of Chimpanzees." *Primates* 8: 355–81.
IZAWA, K. 1970. "Unit Groups of Chimpanzees and Their Nomadism in the Savanna Woodland." *Primates* 11: 1–46.
JOHNSON, R. N. 1972. *Aggression in Man and Animals.* Philadelphia, London.
JOUVET, M. 1972. "Le Discours Biologique." *La Revue de Médicine* 16–17: 1003–63.
KAWAI, M., and MIZUHARA, H. 1959. "An Ecological Study of the Wild Mountain Gorilla." *Primates* 2: 1–42.
KLOPFER, P. H. 1971. "Mother Love: What Turns It On?" *American Scientist* 59: 404–07.
KLUTSCHAK, H. W. 1881. *Als Eskimo unter Eskimos.* Vienna.
KOCH, K. F. 1970. "Cannibalistic Revenge in Jalé Society." *Natural History* 79, 2: 40–51.
———. 1974. *War and Peace in Jalémó: The Management of Conflict in Highland New Guinea.* Cambridge, Mass.
KOCHMAN, T. 1970. "Toward an Ethnography of Black American Speech Behavior." In *Afro-American Anthropology: Contemporary Perspectives*, ed. N. E. Whitten and J. F. Szwed, pp. 145–62. New York.
KOENIG, O. 1970. *Kultur und Verhaltensforschung.* Munich.
KOHL-LARSEN, L. 1943. *Auf den Spuren des Vormenschen.* Stuttgart.
———. 1958. *Wildbeuter in Ostafrika: die Tindiga, ein Jäger- und Sammlervolk.* Berlin.
KONEČNI, V. J. 1975. "The Mediation of Aggressive Behavior: Arousal Level Versus Anger and Cognitive Labeling." *Journal of Personality and Social Psychology* 32: 706–12.
———, and DOOB, A. N. 1972. "Catharsis through Displacement of Aggression." *Journal of Personality and Social Psychology* 23: 379–87.
———, and EBBESEN, E. B. 1976. "Disinhibition vs. the Cathartic Effect: Artifact and Substance." *Journal of Personality and Social Psychology* 34: 352–65.
KÖNIG, H. 1925. "Der Rechtsbruch und sein Ausgleich bei den Eskimos." *Anthropos* 20: 276–315.
KONNER, M. J. 1972. "Aspects of the Developmental Ethology of a Foraging People." In *Ethological Studies of Child Behavior*, ed. N. Blurton-Jones, pp. 285–304.

KORTLANDT, A. 1962. "Chimpanzees in the Wild." *Scientific American*, 206 (5): 128–38.

———. 1963. "Bipedal Armed Fighting in Chimpanzees." *Proceedings of 16th International Congress of Zoologists* 3: 64.

———. 1967. "Handgebrauch bei freilebenden Schimpansen." In *Handgebrauch und Verständigung bei Affen und Frühmenschen*, ed. B. Rensch, 59–102. Berne.

———. 1972. *New Perspectives on Ape and Human Evolution*. Amsterdam.

———, and KOOIJ, M. 1963. "Protohominid Behavior in Primates." *Symposia of the Zoological Society of London* 10: 61–68.

KRÄMER, A. 1902 and 1903. *Die Samoa Inseln. Entwurf einer Monographie unter besonderer Berücksichtigung Deutsch-Samoas*. 2 vols. Stuttgart.

KRAMER, G. 1949. "Macht die Natur Konstruktionsfehler?" *Wilhelmshavener Vorträge, Schriftenreihe der Nordwestdeutschen Universitätsgesellschaft* 1: 1–19.

KRUIJT, J. 1964. "Ontogeny of Social Behaviour in Burmese Red Jungle Fowl (*Gallus gallus spadiceus*)." *Behaviour*, Suppl. 12.

KÜHN, H. 1929. *Kunst und Kultur der Vorzeit. Das Paläolithikum*. Berlin.

———. 1958. *Auf den Spuren des Eiszeitmenschen*. Munich.

KUMMER, H. 1970. "Spacing Mechanisms in Social Behavior." In *Man and Beast: Comparative Social Behavior*, Smithsonian Annual 3, ed. J. F. Eisenberg and W. S. Dillon, pp. 219–34. Washington, D.C.

———. 1973. "Aggression bei Affen." In *Der Mythos vom Aggressionstrieb*, ed. A. Plack, pp. 69–91. Munich.

———, GÖTZ, W., and ANGST, W. 1974. "Triadic Differentiation: An Inhibitory Process Protecting Pair Bonds in Baboons." *Behaviour* 49: 62–87.

LACK, D. 1943. *The Life of the Robin*. Cambridge.

LAGERSPETZ, K. 1969. "Aggression and Aggressiveness in Laboratory Mice." In *Aggressive Behaviour*, ed. S. Garattini and E. B. Sigg, pp. 77–85. Amsterdam.

———. 1974. "Interrelations in Aggression Research: A Synthetic Overview." *Psykologiska Rapporter* 4. Åbo, Finland.

———, and LAGERSPETZ, K. Y. H. 1971 "Changes in the Aggressiveness of Mice Resulting from Selective Breeding, Learning, and Social Isolation." *Scandinavian Journal of Psychology* 12: 241–48.

LANDTMAN, G. 1927. *The Kiwai Papuans of British New Guinea*. London.

LANDY, D., and MATTEE, D. 1969. "Evaluation of an Aggressor as a Function of Exposure to Cartoon Humor." *Journal of Personality and Social Psychology* 9: 237–41.

LAWICK-GOODALL, J. VAN. 1968. "The Behaviour of Free-Living Chimpanzees in the Gombe Stream Area." *Animal Behaviour Monographs*, vol. 1, pt. 3, 161–311.

———. 1971. *In the Shadow of Man*. Boston, London.

———. 1975. "The Behaviour of the Chimpanzee." In *Hominisation und*

Verhalten, ed. G. Kurth and I. Eibl-Eibesfeldt, pp. 74–136. Stuttgart.
LAYARD, J. 1942. *Stone Men of Malekula*. London.
LEAHY, M. J., and CRAIN, M. 1937. *The Land that Time Forgot*. London.
LEAKEY, R. E., and LEWIN, R. 1978. *People of the Lake: Mankind and Its Beginnings*. Garden City.
LEBZELTER, V. 1934. *Eingeborenenkulturen von Süd und Südwestafrika*. Leipzig.
LEE, R. B. 1968. "What Hunters Do for a Living." In *Man the Hunter*, ed. R. B. Lee and I. DeVore, pp. 30–48. Chicago.
———. 1969. "!Kung Bushmen Violence." Paper read at meeting of American Anthropological Association, November 1969.
———. 1972. "!Kung Spatial Organization: An Ecological and Historical Perspective." *Human Ecology* 1: 125–47.
———. 1972. "The !Kung Bushmen of Botswana." In *Hunters and Gatherers Today: A Socioeconomic Study of Eleven Such Cultures in the Twentieth Century*, ed. M. G. Bicchieri, pp. 327–68. New York.
———, and DEVORE, I. 1968. *Man the Hunter*. Chicago.
LEHRMAN, D. S. 1961. "The Presence of the Mate and of Nesting Material as Stimuli for the Development of Incubation Behavior and for Gonadotropic Secretion in the Ring Dove." *Endocrinology* 68: 507–16.
———. 1970. "Semantic and Conceptual Issues in the Nature-Nurture Problem." In *Development and Evolution of Behavior*, ed. L. R. Aronson, E. Tobach, D. S. Lehrman, and J. S. Rosenblatt, pp. 17–52. San Francisco.
LEONG, C. Y. 1969. "The Quantitative Effect of Releasers on the Attack Readiness of the Fish *Haplochromis burtoni* (Cichlidae, Pisces)." *Zeitschrift für vergleichende Physiologie* 65: 29–50.
LEPENIES, W., and NOLTE, H. 1971. *Kritik der Anthropologie*. Munich.
LEYHAUSEN, P. 1973. *Verhaltensstudien an Katzen*. 3d ed. Berlin.
LICHTENSTEIN, H. 1811. *Reisen im südlichen Afrika*. Reprint (1967), vol. 1.
LISCHKE, G. 1972. *Aggression und Aggressionsbewältigung. Theorie und Praxis, Diagnose und Therapie*. Freiburg.
LORENZ, K. 1943. "Die angeborenen Formen möglicher Erfahrung." *Zeitschrift für Tierpsychologie* 5: 235–409.
———. 1953. "Die Entwicklung der vergleichenden Verhaltensforschung in den letzten 12 Jahren." *Zoologischer Anzeiger Supplement* 16: 36–58.
———. 1966. *On Aggression*. New York, London. Translation of (1963) *Das sogenannte Böse*.
———. 1969. "Innate Basis of Learning." In *On the Biology of Learning*, ed. H. Pribram, pp. 13–93. New York.
———. 1970. "Companions as Factors in the Bird's Environment." In *Studies in Animal and Human Behaviour*, 1: 101–258. Cambridge, Mass. Translation of (1935) "Der Kumpan in der Umwelt des Vogels." *Journal für Ornithologie* 83: 137–413.

———1971. "Part and Parcel in Animal and Human Society: A Methodological Discussion." In *Studies in Animal and Human Behaviour,* 2: 115–95. Cambridge, Mass. Translation of (1950) "Ganzheit und Teil in der tierischen und menschlichen Gemeinschaft." *Studium Generale* 3: 455–99.

———. 1971. "Der Mensch, biologisch gesehen. Eine Antwort an Wolfgang Schmidbauer." *Studium Generale* 24: 495–515.

———. 1977. *Behind the Mirror: A Search for a Natural History of Knowledge.* New York. Translation of (1973) *Die Rückseite des Spiegels. Versuch einer Naturgeschichte menschlichen Erkennens.*

LÜERS, F. 1919. "Volkskundliches aus Steinberg beim Achensee in Tirol." *Bayerische Hefte für Volkskunde* 6: 106–30.

LUMHOLTZ, C. 1890. *Among Cannibals. An Account of Four Years' Travels in Australia and of Camp Life with the Aborigines of Queensland.* London.

LUMSDEN, M. 1970. "The Instinct of Aggression: Science or Ideology?" *Futurum: Zeitschrift für Zukunftsforschung* 3: 408–19.

LURIA, S. E. 1973. *Life: The Unfinished Experiment.* New York.

LYNGE, K. 1938–39. *Kalâdlit oqalugtuait oqalualâvilo 1–3.* Nûk.

MC GHEE, P. E. 1972. "On the Cognitive Origins of Incongruity Humor: Fantasy Assimilation versus Reality Assimilation." In *The Psychology of Humor: Theoretical Perspectives and Empirical Issues,* ed. J. H. Goldstein and P. E. McGhee. New York, London.

MACKINTOSH, J. H., and GRANT, E. C. 1966. "The Effect of Olfactory Stimuli on the Agonistic Behaviour of Laboratory Mice." *Zeitschrift für Tierpsychologie* 23: 584–87.

MALLICK, S. K., and MC CANDLESS, B. R. 1966. "A Study of Catharsis of Aggression." *Journal of Personality and Social Psychology* 4: 591–96.

MANING, F. E. 1876. *Old New Zealand: A Tale of the Good Old Times; and a History of the War in the North.* London. Quoted in A. P. Vayda, (1970), "Maoris and Muskets in New Zealand: Disruption of a War System." *Political Science Quarterly* 85: 560–84.

MARCUSE, H. 1969. *Triebstruktur und Gesellschaft.* Frankfurt.

MARKL, H. 1972. "Aggression und Beuteverhalten bei Piranhas (*Serrasalminae*). *Zeitschrift für Tierpsychologie* 30: 190–216.

———. 1974. "Die Evolution des sozialen Lebens der Tiere." In *Verhaltensforschung,* ed. K. Immelmann (Supplementary volume to Bernhard Grzimek, *Tierleben*), pp. 461–85. Munich.

MARLER, P. R., and HAMILTON, W. J. 1966. *Mechanisms of Animal Behavior.* New York, London.

MARSHALL, L. 1959. "Marriage among the !Kung-Bushmen." *Africa* 29: 335–65.

———. 1960. "!Kung-Bushmen Bands." *Africa* 30: 325–55.

———. 1961. "Sharing, Talking, and Giving: Relief of Social Tensions among !Kung-Bushmen." *Africa* 31: 231–49.

———. 1965. "The !Kung-Bushmen of the Kalahari Desert." In *Peoples of Africa,* ed. J. L. Gibbs, pp. 241–78. New York.

MARSHALL, T. E. 1959. *The Harmless People.* London.

MATTHIESSON P. 1962. *Under the Mountain Wall: A Chronicle of Two Seasons in the Stone Age*. New York.
MAUSS, M. 1954. *The Gift: Forms and Functions of Exchange in Archaic Societies*. Glencoe, Ill., London.
MAYNARD-SMITH, J., and PRICE, G. R. 1973. "The Logic of Animal Conflicts." *Nature* 246: 15–18.
MEAD, M. 1928. "The Role of the Individual in Samoan Culture." *Journal of the Royal Anthropological Institute* 58: 481–96.
———. 1928. *Coming of Age in Samoa*. See *From the South Seas*.
———. 1930. *Social Organization of Manua*. B. P. Bishop Museum Bulletin no. 76. Honolulu.
———. 1930. *Growing Up in New Guinea*. See *From the South Seas*.
———. 1935. *Sex and Temperament*. See *From the South Seas*.
———. 1939. *From the South Seas*. Reprint containing *Coming of Age in Samoa, Growing Up in New Guinea*, and *Sex and Temperament*. New York.
———. 1968. "Alternatives to War." In *War: The Anthropology of Armed Conflict and Aggression*, ed. M. Fried, M. Harris, and R. Murphy, pp. 215–28. Garden City, N.Y.
———. 1969. "How Do Children Learn to Govern Their Own Violent Impulses?" *American Journal of Orthopsychiatry* 39: 227–29.
MEGARGEE, E. I., and HOKANSON, J. E. 1970. *The Dynamics of Aggression: Individual, Group, and International Analyses*. New York.
MERZ, F. 1965. "Aggression und Aggressionstrieb." In *Handbuch der Psychologie*, ed. H. Thomae. Vol. 2, *Motivation*, pp. 569–601. Göttingen.
MEVES, C. 1971. *Manipulierte Masslosigkeit*. Freiburg.
MICHAELIS, W. 1974. "Der Aggressions-'Trieb' im Streit der Zoologie und Psychologie." *Naturwissenschaftliche Rundschau* 27: 253–65.
MILGRAM, S. 1963. "Behavioral Study of Obedience." *Journal of Abnormal and Social Psychology* 67: 372–78.
———. 1966. "Einige Bedingungen von Autoritätsgehorsam und seiner Verweigerung." *Zeitschrift für experimentelle und angewandte Psychologie* 13: 433–63.
———. 1974. *Obedience to Authority: An Experimental View*. New York.
MITSCHERLICH, A. 1969. *Die Idee des Friedens und die menschliche Aggressivität*. Frankfurt.
MOHR, A. 1971. "Häufigkeit und Lokalisation von Frakturen und Verletzungen am Skelett vor- und frühgeschichtlicher Menschengruppen." *Ethnographisch-Archäologische Zeitschrift* 12: 139–42.
MONTAGU, A. ed. 1968. *Man and Aggression*. New York.
———. 1973. Discussion remark in *Aggressionstrieb und Krieg*, ed. W. Hollitscher. Stuttgart.
———. 1976. *The Nature of Human Aggression*. New York.
MOREY, R. V., JR., and MARWITT, J. P. 1973. "Ecology, Economy and Warfare in Lowland South America." Paper read at 9th International Congress of Anthropological and Ethnological Sciences, August-September 1973, Chicago.

MORRIS, D. 1967. *The Naked Ape*. London, New York.

———. 1969. *The Human Zoo*. London, New York.

MOYER, K. E. 1968/69. "Internal Impulses to Aggression." *Transactions of the New York Academy of Sciences*, 2d ser., 31: 104–14.

———. 1971. "Experimentale Grundlagen eines physiologischen Modells aggressiven Verhaltens." In *Aggressives Verhalten*, ed. A. Schmidt-Mummendey and H. D. Schmidt. Munich.

———. 1971. *The Physiology of Hostility*. Chicago.

MÜHLMANN, W. E. 1940. *Krieg und Frieden. Ein Leitfaden der politischen Ethnologie*. Heidelberg.

NADEL, S. F. 1951. *The Foundations of Social Anthropology*. Glencoe, Ill., London.

NANCE, J. 1975. *The Gentle Tasaday: A Stone Age People in the Philippine Rain Forest*. New York and London.

NELSON, E. W. 1896/97. *The Eskimo About Bering Strait. Eighteenth Annual Report of the Bureau of American Ethnology*, vol. 1, Washington, D.C.

NESBITT, P. D., and STEVEN, G. 1974. "Personal Space and Stimulus Intensity at a Southern California Amusement Park." *Sociometry* 37: 105–15.

NICE, M. M. 1937. "Studies in the Life History of the Song Sparrow." *Transactions of the Linnaean Society of New York* 4: 57–83.

NISHIDA, T. 1968. "The Social Group of Wild Chimpanzees of the Mahali Mountains." *Primates* 9: 167–224.

———. 1970. "Social Behaviour and Relationship Among Wild Chimpanzees of the Mahali Mountains." *Primates* 11: 47–87.

———, and KAWANAKA, K. 1972. "Inter-Unit Group Relationships among Wild Chimpanzees of the Mahali Mountains." *Kyoto University African Studies* no. 7: 131–69.

NULIGAK. 1961. "Krangmalit . . . Kangeryuarmeut." Nuna, 7, Oblate Mission to the Indians, Cambridge Bay, Northwest Territories, Canada. Quoted in R. Petersen (1963), "Family Ownership and Right of Disposition in Sukkertoppen District, West Greenland." *Folk* 5: 270–81.

NUNGAK, Z., and ARIMA, E. 1969. "Eskimo Stories from Povungnituk." *Bulletin of the National Museum of Canada* no. 235.

OBEREM, U. 1967. "Zur Geschichte des lateinamerikanischen Landarbeiters: Conciertos und Huasipungueros in Ecuador." *Anthropos* 62: 759–88.

OBERHUBER, K. 1972. *Die Kultur des alten Orients. Handbuch der Kulturgeschichte*, pp. 282–87. Frankfurt.

ORTIZ, A. 1969. *The Tewa World: Space, Time, Being, and Becoming in a Pueblo Society*. Chicago.

PACKARD, V. 1962. *The Pyramid Climbers*. New York.

PALLUCK, R. J., and ESSER, A. H. 1971. "Controlled Experimental Modification of Aggressive Behavior in Territories of Severely Retarded Boys." *American Journal of Mental Deficiency* 76: 23–29.

———. 1971. "Territorial Behavior as an Indicator of Changes in Clinical Behavioral Condition of Severely Retarded Boys." *American Journal of Mental Deficiency* 76: 284–90.

PASSARGE, S. 1907. *Die Buschmänner der Kalahari.* Berlin.
PETERSEN, R. 1963. "Family Ownership and Right of Disposition in Sukkertoppen District, West Greenland." *Folk* 5: 270–81.
PETZSCH, H., and PETZSCH, U. 1968. "Neue Beobachtungen zur Fortpflanzungsbiologie von gefangengehaltenen Feldhamstern (*Cricetus cricetus L.*) und daraus ableitbarer Schlussfolgerungen für die Angewandte Zoologie." *Der Zoologische Garten* 35: 256–69.
PHILLIPS, L. H., and KONISHI, M. 1972. "Control of Aggression by Singing in Crickets." *Nature* 241: 64–65.
PITCAIRN, T. K., and SCHLEIDT, M. 1976. "Dance and Decision: An Analysis of a Courtship Dance of the Medlpa, New Guinea." *Behaviour* 58: 298–316.
PITELKA, F. A. 1959. "Numbers, Breeding Schedule and Territory in Pectoral Sandpipers of Northern Alaska." *Condor* 61: 233–64.
PLACK, A. 1968. *Die Gesellschaft und das Böse.* 2d ed. Munich.
———, ed. 1973. *Der Mythos vom Aggressionstrieb.* Munich.
PLOOG, D. W., BLITZ, J., and PLOOG, F. 1963. "Studies on Social and Sexual Behavior of the Squirrel Monkey (*Saimiri sciureus*)." *Folia Primatologica* 1:29–66.
PLOTNIK, R. 1974. "Brain Stimulation and Aggression: Monkeys, Apes and Humans." In *Primate Aggression, Territoriality, and Xenophobia: A Comparative Perspective*, ed. R. L. Holloway, pp. 389–415. New York, London.
POPPER, K. 1961. *The Logic of Scientific Discovery.* New York.
PORTISCH, H. 1970. *Friede durch Angst: Augenzeuge in den Arsenalen des Atomkrieges.* Vienna.
RADCLIFFE-BROWN, A. R. 1933. *The Andaman Islanders: A Study in Social Anthropology.* 2d ed. Cambridge, New York.
———. 1940. "On Joking-Relationships." *Africa* 13: 195–210.
RAPPAPORT, R. A. 1968. *Pigs for the Ancestors: Ritual in the Ecology of a New Guinea People.* New Haven, London.
RASA, O. A. E. 1969. "The Effect of Pair Isolation on Reproductive Success in *Etroplus maculatus* (Cichlidae)." *Zeitschrift für Tierpsychologie* 26: 846–52.
———. 1971. "Appetence for Aggression in Juvenile Damsel Fish." *Zeitschrift für Tierpsychologie*, Supplement 7.
RASMUSSEN, K. 1905. *Nye Mennesker.* Copenhagen.
———. 1908. *People of the Polar North.* London.
———. 1924. *Myter og Saga fra Grønland.* Copenhagen.
———. 1930. *Report of the Fifth Thule Expedition.* Vol. 7, *Intellectual Culture of the Caribou Eskimos.* Copenhagen.
———. 1932. *Report of the Fifth Thule Expedition.* Vol. 9, *Intellectual Culture of the Copper Eskimos.* Copenhagen.
RATTNER, J. 1970. *Aggression und menschliche Natur.* Olten.
REIS, D. J. 1974. "Central Neurotransmitters in Aggression." In *Aggression*, ed. S. H. Frazier. Proceedings of the Association, Research Publications of the Association for Nervous and Mental Diseases, p. 119–48. Baltimore.

REUCK, A. DE, and KNIGHT, J. eds. 1966. *Conflict in Society*. Ciba Foundation Volume. London.

REYER, H. U. 1975. "Ursachen und Konsequenzen von Aggressivität bei *Etroplus maculatus* (Cichlidae, Pisces): Ein Beitrag zum Triebproblem." *Zeitschrift für Tierpsychologie* 39: 415–54.

REYNOLDS, V. 1966. "Open Groups in Human Evolution." *Man* 1: 441–52.

———, and LUSCOMBE, O. 1969. "Chimpanzee Rank Orders and the Function of Displays." *Proceedings of 2d International Congress on Primatology, Zurich 1968*. Basel.

———, and REYNOLDS, F. 1965. "Chimpanzees of the Budongo Forest." In *Primate Behavior: Field Studies of Monkeys and Apes*, ed. I. DeVore, pp. 368–424. New York.

ROEDER, K. D. 1955. "Spontaneous Activity and Behavior." *Scientific Monthly* 80: 362–70.

ROGERS, E. S. 1969. "Band Organization Among the Indians of Eastern Subarctic Canada." *Bulletin of the National Museum of Canada* no. 288, pp. 21–55.

ROPER, M. K. 1969. "A Survey of Evidence for Intrahuman Killing in the Pleistocene." *Current Anthropology* 10: 427–59.

ROTH, G. 1974. "Die verhaltensphysiologischen Grundlagen bei Lorenz." In *Kritik der Verhaltensforschung*, ed. idem, pp. 156–89. Munich.

ROTHBART, M. K. 1973. "Laughter in Young Children." *Psychological Bulletin* 80: 247–56.

RUSSELL, C., and RUSSELL, W. M. S. 1968. *Violence, Monkeys and Man*. London, New York.

SAHLINS, M. D. 1960. "The Origin of Society." *Scientific American* 204: 76–87.

———. 1972. *Stone Age Economics*. Chicago.

SBRZESNY, H. 1976. "Die Spiele der !Ko-Buschleute unter besonderer Berücksichtigung ihrer sozialisierenden und gruppenbindenden Funktionen." *Monographien zur Humanethologie* 2. Munich-Zurich.

SCHALLER, G. B. 1963. *The Mountain Gorilla: Ecology and Behavior*. Chicago.

———. 1972. *The Serengeti Lion. A Study of Predator-Prey Relations*. Chicago.

SCHEBESTA, P. 1938. *Die Bambuti-Pygmäen von Ituri*. Mémoires de l'Institut Royal Colonial Belge, Section des Sciences Morales et Politiques, vol. 1. Brussels.

———. 1941. *Die Bambuti-Pygmäen vom Ituri*. Mémoires de l' Institut Royal Colonial Belge, Section des Sciences Morales et Politiques, vol. 2. Brussels.

———. 1948. *Die Bambuti-Pygmäen vom Ituri*. Mémoires de l' Institut Royal Colonial Belge, Section des Sciences Morales et Politiques, vol. 2. Brussels.

———. 1950. *Die Bambuti-Pygmäen vom Ituri*. Mémoires de l' Institut Royal Colonial Belge, Section des Sciences Morales et Politiques, vol. 2. Brussels.

———. 1951/53. "Über die Vorgeschichte der Völkerschaften von Südwestafrika, pt. 1, Die Büschmänner." *South Africa Journal of Science* 9: 45–56.

SCHINDLER, H. 1974. "Territorialität und Aggression: Eine Erwiderung." *Anthropos* 69: 275–78.

SCHJELDERUP, H. 1963. *Einführung in die Psychologie*. Bern.

SCHJELDERUP-EBBE, T. 1922. "Beiträge zur Sozialpsychologie des Haushuhns." *Zeitschrift für Psychologie* 88: 225–52.

———. 1922. "Soziale Verhältnisse bei Vögeln." *Zeitschrift für Psychologie* 90:106–107.

SCHMIDBAUER, W. 1971. "Methodenprobleme der Humanethologie." *Studium Generale* 24: 462–522.

———. 1971. "Zur Anthropologie der Aggression." *Dynamische Psychiatrie/Dynamic Psychiatry* 4: 36–50.

———. 1972. *Die sogenannte Aggression*. Hamburg.

———. 1973. "Territorialität und Aggression bei Jägern und Sammlern." *Anthropos* 68: 548–58.

———. 1973. *Biologie und Ideologie: Kritik der Humanethologie*. Hamburg.

———, ed. 1974. *Evolutionstheorie und Verhaltensforschung*. Hamburg.

SCHMIDT-MUMMENDEY, A., and SCHMIDT, H. D. 1971. *Aggressives Verhalten*. Munich.

SCHOECK, H. 1966. *Der Neid*. Freiburg, Munich.

SCHULTZE-WESTRUM, T. 1974. *Biologie des Friedens. Auf dem Weg zu solidarischem Verhalten*. Munich.

SCHUMACHER, P. 1950. *Kivu-Pygmäen*. Mémoires de l' Institut Royal Colonial Belge, Section des Sciences Morales et Politiques, vol. 2. Brussels.

SCOTT, J. P. 1958. *Aggression*. Chicago.

SELG, H. 1968. *Diagnostik der Aggressivität*. Göttingen.

———. 1971. *Zur Aggression verdammt?*. Stuttgart.

SERVICE, E. R. 1962. *Primitive Social Organization: An Evolutionary Perspective*. New York.

SHERIF, M., and SHERIF, C. W. 1966. *Groups in Harmony and Tension*. New York.

SHIRER, W. L. 1960. *The Rise and Fall of the Third Reich*. New York.

SHOKEID, M. 1973. "Conflict and Entertainment: An Analysis of Social Gatherings and Celebrations Among Moroccan Immigrants in Israel." Paper read at 9th International Congress of Anthropological and Ethnological Sciences, August-September 1973, Chicago.

SILBERBAUER, G. P. 1972. "The G/wi Bushmen." In *Hunters and Gatherers Today: A Socioeconomic Study of Eleven Such Cultures in the Twentieth Century*, ed. M. G. Bicchieri, pp. 271–326. New York.

———. 1973. "Socio-Ecology of the G/wi Bushmen." Thesis, Department of Anthropology and Sociology, Monash University.

SINGER, D. 1968. "Aggression Arousal, Hostile Humor, Catharsis." *Journal of Personality and Social Psychology*, Monograph Supplement, 8, vol. 1, pt. 2.

SIPES, R. G. 1973. "War, Sports and Aggression: An Empirical Test of Two Rival Theories." *American Anthropologist* 75: 64–86.

SKINNER, B. F. 1971. *Beyond Freedom and Dignity*. New York.

SPERRY, R. W. 1971. "How a Developing Brain Gets Itself Properly Wired for Adaptive Function." In *The Biopsychology of Development*, ed. E. Tobach, L. R. Aronson, and E. Shaw, pp. 27–44. New York, London.

SPITZ, R. 1950. "Anxiety in Infancy." *International Journal of Psycho-Analysis* 31: 139–43.

———, and COBLINER, W. G. 1965. *The First Year of Life, A Psychoanalytic Study of Normal and Deviant Development of Object Relations*. New York.

SROUFE, L. A. 1977. "Wariness of Stranger and the Study of Infant Development." *Child Development* 48: 731–46.

STAYTON, D. J., HOGAN, R., and AINSWORTH, M. D. S. 1971. "Infant Obedience and Maternal Behavior: The Origins of Socialization Reconsidered." *Child Development* 42: 1057–69.

STRATHERN, A. 1971. *The Rope of Moka: Big-Men and Ceremonial Exchange in Mount Hagen, New Guinea*. London, New York.

STREHLOW, C. 1915. "Die Aranda- und Loritja-Stämme in Central-Australien." *Veröffentlichungen des Städtischen Völkermuseums Frankfurt/M*, pt. 4, sec. 2, pp. 1–78.

SUGIYAMA, Y. 1969. "Social Behavior of Chimpanzees in the Budongo Forest, Uganda." *Primates* 10: 197–225.

SWEET, W. H., ERVIN, F., and MARK, V. H. 1969. "The Relationship of Violent Behaviour to Focal Cerebral Disease." In *Aggressive Behaviour*, ed. S. Garattini and E. B. Sigg, pp. 336–52. Amsterdam.

TELEKI, G. 1973. *The Predatory Behavior of Wild Chimpanzees*. Lewisburg, Pa.

TELLEGEN, A., HORN, J. M., and LEGRAND, R. G. 1969. "Opportunity for Aggression as Reinforcer in Mice," *Psychonomic Science* 14: 104–105.

THOMAS, E. M., 1959. *The Harmless People*. New York, London.

THOMPSON, T. I. 1963. "Visual Reinforcement in Siamese Fighting Fish." *Science* 141: 55–57.

———. 1964. "Visual Reinforcement in Fighting Cocks." *Journal of Experimental Analysis of Behavior* 7: 45–49.

TINBERGEN, E. A., and TINBERGEN, N. 1972. "Early Childhood Autism—An Ethological Approach." *Zeitschrift für Tierpsychologie*, Supplement 10.

TINBERGEN, N. 1951. *The Study of Instinct*. London, Toronto.

———. 1968. "On War and Peace in Animals and Man." *Science* 160: 1411–18.

TOBACH, E., GIANUTSOS, J., TOPOFF, H. R., and GROSS, C. G. 1974. *The Four Horsemen: Racism, Sexism, Militarism and Social Darwinism*. New York.

TOBIAS, P. VON 1964. "Bushman Hunter-Gatherers: A Study in Human Ecology." In *Ecological Studies in Southern Africa*, ed. D. H. S. Davis. The Hague. Reprinted (1968) in *Man in Adaptation: The Cultural Present*, ed. Y. A. Cohen, pp. 196–208. Chicago.

TOMPA, F. S. 1962. "Territorial Behaviour: The Main Factor Controlling a Local Song Sparrow Population." *Auk* 79: 687–97.

TOYNBEE, A. 1966. "Tradition und Instinkt." In *Vom Sinn der Tradition*, ed. L. Reinisch, pp. 35–52. Munich.

TRENK, [OBERLEUTNANT]. 1910. "Die Buschleute der Namib, ihre Rechts- und Familienverhältnisse." *Mitteilungen der deutschen Schutzgebiete* 23: 166–70.

TRIVERS, R. L. 1971. "The Evolution of Reciprocal Altruism." *Quarterly Review of Biology* 46: 35–57.

TURNBULL, C. M. 1961. *The Forest People*. London, New York.

———. 1965. "The Mbuti Pygmies: An Ethnographic Survey." *American Museum of Natural History Anthropological Papers* no. 50 (3): 139–282.

———. 1968. "The Importance of Flux in Two Hunting Societies." In *Man the Hunter*, ed. R. B. Lee and I. DeVore, pp. 132–37. Chicago.

TURNEY-HIGH, H. H. 1949. *Primitive War: Its Practice and Concepts.* Columbia, S.C.

VALLOIS, H. V. 1961. "The Social Life of Early Man: The Evidence of Skeletons." In *Social Life of Early Man*, ed. S. L. Washburn, pp. 214–35. Chicago.

VALZELLI, L. 1969. "Aggressive Behaviour Induced by Isolation." In *Aggressive Behaviour*, ed. S. Garattini and E. B. Sigg, pp. 70–76. Amsterdam.

VAYDA, A. P. 1960. "Maori Warfare." In *Polynesian Society Maori Monographs* no. 2. Wellington.

———. 1961. "Expansion and Warfare Among Swidden Agriculturalists." *American Anthropologist* 63: 346–58.

———. 1967. "Research on the Functions of Primitive War." In *Peace Research Society International Papers*, vol. 7.

———. 1968. "Hypotheses About Functions of War." In *War: The Anthropology of Armed Conflict and Aggression*, ed. M. Fried, M. Harris, and R. Murphy, pp. 85–91. Garden City, N.Y.

———. 1970. "Maoris and Muskets in New Zealand: Disruption of a War System." *Political Science Quarterly* 85: 560–84.

VEDDER, H. 1937. "Die Buschmänner Südwestafrikas und ihre Weltanschauung." *South African Journal of Science* 24: 416–36.

———. 1952/53. "Über die Vorgeschichte der Völkerschaften von Südwestafrika, pt. 1, Die Buschmänner." *Journal of the South-West Africa Scientific Society* 9: 45–56.

VICEDOM, G. F. 1937. "Ein neuentdecktes Volk in Neuguinea." *Archiv für Anthropologie*, n.s. 24: 11–14.

———, and TISCHNER, H. 1943/48. *Die Kultur der Hagenbergstämme*, vol. 1, *Die Mbowamb*. Monographien zur Völkerkunde no. 1. Hamburg.

VOLHARD, E. 1939. *Kannibalismus*. Stuttgart. Reprinted (1968) New York.

WAAL, F. M. B. DE. 1978. "Exploitative and Familiarity Dependent Support Strategies in Chimpanzees." *Behaviour* 17: 268–312.

WALSH, M. N. 1971. "Psychic Factors in the Causation of Recurrent Mass Homicide." In *War and the Human Race*, ed. M. N. Walsh, pp. 70–82. New York.

WALTERS, R. H., and THOMAS, E. L. 1963. "Enhancement of Punitiveness by Visual and Audiovisual Displays." *Canadian Journal of Psychology* 17: 244–55.

WALTHER, F. R. 1961. "Entwicklungszüge im Kampf- und Paarungsverhalten der Horntiere." *Jahrbuch Georg von Opel Freigehege für Tierforschung* 3: 90–115.

———. 1966. *Mit Horn und liuf*. Berlin.

WARNER, W. L. 1930. "Murngin Warfare." *Oceania* 1: 457–82. v494.

WEIDKUHN, P. 1968/69. "Aggressivität und Normativität. Über die Vermittlerrolle der Religion zwischen Herrschaft und Freiheit. Ansätze zu einer kulturanthropologischen Theorie der sozialen Norm." *Anthropologie* 63/64: 361–94.

———. 1969. "Fastnacht–Revolte–Revolution." *Zeitschrift für Religion und Geistesgeschichte* 21: 289–306.

WEULE, K. 1916. *Der Krieg in den Tiefen der Menschheit*. Stuttgart.

WICKLER, W. 1965. "Über den taxonomischen Wert homologer Verhaltensmerkmale." *Die Naturwissenschaften* 52: 441–44.

———. 1966. "Ursprung und biologische Deutung des Genitalpräsentierens männlicher Primaten." *Zeitschrift für Tierpsychologie* 23: 422–37.

———. 1967. "Vergleichende Verhaltensforschung und Phylogenetik." In *Die Evolution der Organismen*, 3d ed. vol. 1, ed. G. Heberer, pp. 420–508. Stuttgart.

WICKLER, W. 1969. *Sind wir Sünder? Naturgesetze der Ehe*, Munich.

———. 1970. "Soziales Verhalten als ökologische Anpassung." In *Verhandlungen der Deutschen Zoologischen Gesellschaft, 64. Jahresversammlung 1970*, ed. W. Rathmayer, pp. 291–304. Stuttgart.

———. 1971. *Die Biologie der Zehn Gebote*. Munich.

———. 1972. *Verhalten und Umwelt*. Hamburg.

WILHELM, J. H. 1953. "Die !Kung-Buschleute." *Jahrbuch des Museums für Völkerkunde Leipzig*, 12: 91–189.

WILLIS, E. O. 1967. *The Behavior of the Bicolored Antbird*. University of California Publications in Zoology no. 79. Berkeley.

WILMSEN, E. N. 1973. "Interaction, Spacing Behavior, and the Organization of Hunting Bands." *Journal of Anthropological Research* 29: 1–31.

WILSON, E. O. 1971. "Competitive and Aggressive Behavior." In *Man and Beast: Comparative Social Behavior*. Smithsonian Annual 3, ed. J. F. Eisenberg and W. S. Dillon, pp. 183–217. Washington, D.C.

———. 1974. "Wettbewerb und Aggression bei Tier und Mensch." In *Evolutionstheorie und Verhaltensforschung*, ed. W. Schmidbauer, pp. 259–94. Hamburg.

———. 1975. *Sociobiology, the New Synthesis*. Cambridge, Mass.

WINTSCH, H. U. 1973. "Erziehung zur Friedfertigkeit." In *Mythos vom Aggressionstrieb*, ed. A. Plack, pp. 285–310. Munich.

WOODBURN, J. 1968. "Stability and Flexibility in Hadza Residential Groupings." In *Man the Hunter*, ed. R. B. Lee and I. DeVore, pp. 103–10. Chicago.

WRIGHT, Q. 1965. *A Study of War*, 2d ed. Chicago.

YOUNG. M. V. 1971. *Fighting with Food—Leadership, Values and Social Control in a Massim Society*. Cambridge.

ZASTROW, B. VON, and VEDDER, H. 1930. "Die Buschmänner." In *Das Eingeborenenrecht: Togo, Kamerun, Südwestafrika, die Südseekolonien*, ed. E. Schultz-Ewerth and L. Adam. Stuttgart.

Films of the Human Ethology Film Library of the Max Planck Society Mentioned in the Text

EIBL-EIBESFELDT, I. 1971. "!Ko Bushmen (Kalahari)—Mocking and Genital Display" (Humanethologisches Filmarchiv HF 1). *Homo* 22: 261–66.

———. 1971. "!Ko Bushmen (Kalahari)—Aggressive Behavior of Children of Prepubertal Age, pts. 1 and 2" (Humanethologisches Filmarchiv HF 2 and HF 3). *Homo* 22: 267–78.

———. 1972. "!Ko Bushmen (Kalahari)—Aggressive Behavior of Babies" (Humanethologisches Filmarchiv HF 4). *Homo* 23: 292–97.

———. 1974. "!Kung Bushmen (Kungveld, South-West Africa)—Sibling Rivalry, Mother-Child Interactions" (Humanethologisches Filmarchiv HF 41). *Homo* 24: 252–60.

———. 1979. "*Medlpa* (New Guinea)—Mourning" (Humanethologisches Filmarchiv HF 83).

Other film publications of this library are mentioned in Eibl-Eibesfeldt *Die !Ko-Buschmanngesellschaft, Gruppenbindung und Aggressionskontrolle* (Munich, 1972) and *Der vorprogrammierte Mensch* (Vienna, 1973).

Index